INTO THE DREAM LAB

INTO THE DREAM LAB

The New Science of Dreams and Nightmares

MICHELLE CARR

Profile Books

First published in Great Britain in 2025 by
Profile Books Ltd
29 Cloth Fair
London
EC1A 7JQ

www.profilebooks.com

Copyright © Michelle Carr, 2025

1 3 5 7 9 10 8 6 4 2

Typeset in Berling Nova Text by MacGuru Ltd
Printed and bound in Great Britain by
CPI Group (UK) Ltd, Croydon CR0 4YY

The moral right of the author has been asserted.

All rights reserved. Without limiting the rights under copyright reserved above, no part of this publication may be reproduced, stored or introduced into a retrieval system, or transmitted, in any form or by any means (electronic, mechanical, photocopying, recording or otherwise), without the prior written permission of both the copyright owner and the publisher of this book.

A CIP catalogue record for this book is
available from the British Library.

Our product safety representative in the EU is Authorised Rep Compliance Ltd., Ground Floor, 71 Lower Baggot Street, Dublin, D02 P593, Ireland. www.arccompliance.com

ISBN 978 1 80522 028 2
eISBN 978 1 80522 030 5

For the days and nights of reverie in the
Dream and Nightmare Lab

I believe that the nightmare, far from being a failed or aberrant dream, is one of the most important kinds of dream, and the one in which we can most easily observe a process which probably occurs in all dreams. In this sense the nightmare is the most useful of dreams.

– Dr Ernest Hartmann

Contents

Introduction   1

Part I: How Dreams Work
1. The Scaffolds of the Dream World   13
2. The Dreaming Brain   52

Part II: Why Dreams and Nightmares Matter
3. Why Dream at All?   89
4. When Are Nightmares a Problem?   123

Part III: Working with Dreams and Nightmares
5. Treating Nightmares and Going Lucid   157
6. Engineering Dreams   193

Part IV: Where Else Is Dreaming Relevant?
7. Bad Dreams and Health   233
8. Sleep On It: Dream Skills   270

Conclusion   306

Notes   315
Acknowledgements   330
Index   333

Introduction

It was just a dream.

Most people think of nightmares as something to forget about, to leave behind in the dark of night. Something we have no control over.

But pretty much all of us have bad dreams. We don't often talk about them, preferring to shrug them off in the morning light. But bad dreams are about much more than just a poor night's sleep, and understanding the impact – and surprising upside – to these dreams can help to soothe our restless nights and support healthy minds in waking life.

It's normal to recall bad dreams during periods of stress and, in some ways, it seems like these dreams help us to deal with difficult events from the day – what some scientists have described as 'overnight therapy'. Others see bad dreams as an evolutionary gain, a safe space where we can practise responding to potential real-life threatening scenarios. Most scientists agree that there is an upside to bad dreams, an advantage to facing these inner demons at night.

At the extreme, though, chronic nightmares can seep into our waking lives and be harmful to mental health.

As an example, in a sample of over fifteen thousand hospital workers, of those exposed to trauma, experiencing frequent nightmares was the single greatest risk factor for suicide.

Why is this? How can something so apparently illusory have such a grave impact on our psyche?

What most people don't realise is this: A nightmare is a real experience. It is real in our mind, in our brain, and in our body. For people who have faced significant trauma, or even those who are highly sensitive to stress, nightmares can in essence retraumatise an individual each time they occur. On top of this, frequent nightmares, and the disturbed sleep they cause, can interfere with our brain's ability to regulate emotion so we are left more defenceless the next day.

Luckily, modern dream science is drawing back the curtain on the dreaming mind, revealing what's going on inside after we turn off the lights. And cutting-edge research is giving us new ways to interface with our bodies and minds at rest, to influence these hidden worlds we inhabit when we sleep, and to stop the disruptive cycle of nightmares and restore a good night's slumber.

I'm a dream scientist. Over the past two decades I've spent hundreds of nights awake in the laboratory, watching people sleep. I watch their brain waves on the monitor, growing slower as they fall into deep sleep. Around every ninety minutes, the brain practically wakes up, and their eyes dart around as if watching a scene, or likely a vivid dream. I wake them lightly through the speaker: 'Can you tell me what you were dreaming about?'

Introduction

Much of my work has focused on nightmares and what happens in the mind, brain and body when they occur. If someone has a nightmare in the lab, we might see their heart racing, their breathing becoming erratic and their brain activity fast and prone to waking. Their minds and bodies are convinced that the fear is real. And in many ways, it is. The real stress of nightmares often spills into waking life, with harmful repercussions, especially to mental health. Thus, their study and treatment are vital areas of research.

There are three main questions at the heart of my work: Why do we dream? Why do dreams go bad? And how can we harness the science of dreams and nightmares to improve our health? In this book, you'll learn about the evolutionary mechanisms behind the purpose of dreaming. You'll learn about the science behind when things go wrong and the possibilities that understanding nightmares offers. And you'll learn how to use dreams in your own life, either therapeutically or just for fun – and even how to lucid dream, too.

To study nightmares in the dream lab, we place various sensors on our sleeping subjects: Electrodes on the scalp measure brain activity to allow us to determine sleep stages; electrodes on the body measure heart rate and muscle tension, including two above the eyebrow – if you notice someone frowning during sleep, they could be experiencing a nightmare. An infrared camera projects a video of the sleeper into the control room, so experimenters can observe them over the course of the night. Periodically throughout sleep, we wake people up to recount their dreams.

While it's common knowledge that we spend a good portion of our lives asleep – eight hours each night, or a third of each day – what most people don't realise is that a lot of our time in sleep is spent dreaming. When I wake subjects repeatedly throughout the night – from the first moments of sleep onset through the depths of deep sleep and into the light hours of morning – in most cases they can remember that just before awakening they were experiencing something – thinking, feeling, dreaming something. Their sleeping mind is remarkably active; it slumbers but does not cease.

The content of these dreams varies across the night, at times more mundane and thought-like, at others more elaborate and dreamlike. In a way, this is similar to how your thoughts change as you go about your daily life: mind-wandering while doing the dishes or commuting to work, mind-focused while conversing with a friend or putting together a bookshelf. The varied inner experiences we have make up the phenomenal quality of our life – both while waking and while sleeping – and even define our mental health over time. It's not just the behaviours or tasks of the day that dictate our well-being, or even the state of our biology. Our inner experience of life is vital to our health and, as we'll soon see, how well we sleep at night.

The mind that you carry into sleep matters. We continue to have mental experiences across the night – thoughts and feelings, sensory impressions and emotions – and this activity of the sleeping mind is directly tied to our health. When dreams are agitated and distressing, they fill the body with real stress that spills over into waking life. And the disruption of sleep by nightmares wears down the

Introduction

brain's defences and makes us more vulnerable the next day; this can lead to an inescapable downward spiral as this distressed mental state is carried into dreaming and back out of it again.

So while the past few decades have firmly established the importance of sleep for all facets of waking health, it has become increasingly clear that dreaming, too, impacts not only the quality of our sleep but also how well we feel and perform each day. Dreams are consequential. How we think, feel and behave in the hidden lives of our sleeping mind – it counts.

You may be wondering, what can we actually do about any of this? Most people hardly hope to exert any kind of control over their dreams. In fact, this belief that we *cannot* control our dreams is so widely entrenched that even people suffering from debilitating nightmares rarely, if ever, seek treatment.

But new understanding of dreams and nightmares says otherwise: As it turns out, we are not destined to accept whatever the night throws at us. Today, research is revealing that dreams are far more malleable and responsive to attention than you may have ever imagined, that we can to some extent engineer our dreams, shaping these inner worlds that seem so beyond our control. The purpose of this book, and of my research in general, is to help people engage purposefully with dreaming, to overcome nightmares, and to have more agency in how we go about our sleeping lives.

Before we get to precisely *how* we can engineer dreams, we must first explore why we dream in the first place. How do our brains and bodies produce and respond to these nightly visions? What are the evolutionary functions of dreams? And what goes wrong in the case of nightmares?

To begin with, we can think of dreaming as a learned behaviour. We learn how to dream as individuals over the course of our lives; we develop personal dream habits and tend to have our own recurring dream themes. We have also learned how to dream as a species. Research has uncovered the same universal themes reappearing in dreams across cultures, suggesting certain dream features have evolved by design, in some way helping us to survive as a species over the course of history.

We also learn *from* our dreams. We might learn how to react when faced with a fear: Do we run and hide? Do we succumb to or overcome a given threat? When caught in a conflict, do we trust other people? Do we turn towards them or away? Dreams like these seem to prime our behaviours the next day, perhaps rousing intimacy or inciting jealousy, or generally affecting our mood. The mind learns from its trials during sleep and translates these lessons to waking life.

And of course, dreams often reflect our waking concerns, what Freud called the 'day residue,'[1] where recent events become swiftly woven into dreams. Most scientists think this process is meaningful, not random – that a natural process of dreaming is to echo through salient moments of the day, helping us to organise our thoughts and worries and adjust to new challenges in life.

In some ways it seems like nightmares interrupt this process, that rather than sorting through the day's events fluently, the dream confronts an obstacle, an

Introduction

insurmountable stress that leads to overwhelm and awakening. In our research we have found that the brains of nightmare sufferers are less able to manage – or regulate – negative emotions when faced with stress. This creates a vicious cycle, where the mind becomes overwhelmed by stress during the day, and this leads to nightmares that further disrupt sleep and cause more distress the next day.

These nightmares, like dreams, seem to be learned. They usually have recurring themes that can reappear over many years, decades even, when triggered by similar waking life events. And most nightmare sufferers feel helpless, unable to do anything about their bad dreams. But it is possible to treat nightmares, to learn new patterns of dreaming in their place. Many nightmare treatments use simple visualisation exercises during the day to work the imaginative muscles while awake, creating new positive themes for dreaming.

Another powerful tool to overcome nightmares is lucid dreaming, where you become aware that you are dreaming while still asleep and can rewrite negative dreams as they occur.

My own path to the science of dreaming began after I experienced my first ever lucid dream. In 2008, I was an undergraduate at the University of Rochester, working in a sleep lab and studying cognitive neuroscience – that mysterious link between the mind and brain. One morning I was struggling to rouse myself from a bad dream, and when I opened my eyes and groggily sat up, I saw that the rest of my body was still lying there asleep! I realised I was dreaming, and marvelled at how real it felt.

After I woke up, I was captivated by the experience.

How was the dreaming mind able to procure such a detailed living experience of the world, while the brain and body were deeply asleep? Over time and through personal experience, I learned that lucid dreaming can be used to overcome nightmares, and in the years since, I became especially interested in the neuroscience of nightmares and lucid dreams, and how we can use these dreams for good.

Today, I'm a professor of dream and nightmare neuroscience, and I direct a laboratory devoted to engineering dreams, to shaping these inner worlds in sleep. I see the waking world opening up to the science of dreams, through the sharpening lens of neurocognitive science, growing networks of laboratories devoted to the study of the sleeping mind, and increasing attention to nightmares in sleep and psychiatric medicine. With this momentum, I hope to shine a light on nightmares and to unveil the myriad techniques available to work with dreams and improve our health in turn.

In this book, we'll take a deep dive into the science behind nightmares and nightmare treatment, explore how to have lucid dreams, and discover new tools for dream engineering that are designed to interface directly with the sleeping mind.

In Part I we will encounter the science of dreaming, uncovering what types of memories, emotions and actions are reflected in dream content, and what functions they serve. We'll examine the peculiar scaffolds of the dream world and surf the waves of the sleeping brain, where patterns of electrical activity drive the generation of dreaming. We will also see that dreaming is not confined

Introduction

to the brain; the content of dreams is linked to the sleeping body and even receptive to the outside world, sensing stimuli from the environment and absorbing these sensations into dreaming.

In Part II we will look at the experience of dreaming. What is it like to be in a dream? How do our feelings and sensations arise through dream content? We will also explore how dreams become amplified in the case of nightmares; how stress and trauma give rise to nightmares, and the consequences on health over time. At the same time, we will see that many people who are prone to nightmares are also highly sensitive, and that this can be valuable at times: nightmare sufferers tend to be highly perceptive, creative and empathetic. This is an often-overlooked upside to being nightmare-prone.

In Part III we will explore how to work with nightmares, first in waking visualisations and then in lucid dreams, to foster more healthy dreaming lives. We'll discover techniques for interfacing directly with dreams and nightmares as they occur. Our lab uses simple lights and sounds to pierce the veil of dreams and enhance lucidity, but others use audio prompts to direct sleep onset dreams, or odours to ameliorate dream quality. Dream engineers are learning to modulate the sleeping mind, to dampen negative emotions or interrupt the bodily stress of nightmares and thereby repair the quality of sleep.

Finally, in Part IV, we will see that even beyond nightmares, dreaming acts as a useful barometer for our sleep and mental health, and a valuable resource we can tap into for a healthier life. Dreaming offers a practical means for building social connection and empathy, harvesting insight and creativity, and strengthening learning. I'll show you how to engage with your dreams through dreamwork, lucid dreaming and dream engineering, and explain how these practices can become vital components to building a healthier sleeping and waking life.

In the end, I hope to bring you to a new understanding of how we experience sleep, why nightmares matter and what you can do to dream well tonight.

Part I

How Dreams Work

1
The Scaffolds of the Dream World

Anyone who has ever dreamed will know that many of the dreams we have are built on the scaffolding of our everyday life. Faces, places, or actions that we're familiar with pop up in our sleeping lives, along with concerns and affairs borrowed from waking reality: an argument or conversation, a regular commute, a movie we've just seen.

After many years of studying the neuroscience of dreams, reading thousands of dream reports collected in the laboratory or at home, through online surveys or in everyday conversation, one thing that has become clear to me is that there are certain patterns in dreaming – patterns underlying how our waking life is not only represented but *mis*represented in the sleeping mind. The dream has a unique, purposeful design: reflecting, yet transforming, our daily lives.

What we pay attention to while awake permeates where we journey in our dreams, with personal memories and culture filling in the scaffolds of the dream world. The sleeping self maintains a sense of identity, too, preoccupied by work, romance, survival. And the external world seeps into dreaming; this will come as no surprise to anyone who has slept through their alarm clock before, the sound disguised as songbirds or disco-themed dreams. Indeed, the sleeping brain is not so cut off from the sensory world, the lights and sounds of the room around, as scientists once believed. All of these components together – our memories and preoccupations and the guise of the sleeping body – drive the design of dreams.

What are the idiosyncratic and, perhaps more telling, universal ways that we dream?

How dreams are designed

Over the course of my career, I've had the privilege of conducting dream science in a number of sleep laboratories around the world. The quintessential sleep lab is built on the same basic principles: a bed in a closed room, a two-way communication system, ideally a video monitor, and various recording equipment. This includes electrodes to measure brain activity and fluctuating heart rate, a small tube that splits into the nostrils to measure airflow, and various electrodes measuring twitches or tension in muscles or eye movements. Even more sensors can measure the position of the head and body, or the rising and falling of the chest, and a small microphone taped to the neck can record the slightest tickle of snoring in the throat. All of these subtle signals of the living body are collected in massive amounts of data over an eight-hour night of sleep.

Countless times I've invited subjects into the laboratory, and before they go to sleep I spend some time placing electrodes on their scalp and casually making small talk. I try keeping the lights low, offering some herbal tea, and

The Scaffolds of the Dream World

putting the subject at ease: the human element is essential to a good night's sleep.

Invariably, almost every subject asks me how I got into this work. It's a fair question – dream scientist is hardly a typical job title. And while I have always been fascinated by dreams, I must admit that I never thought I would be able to pursue a career in the science of dreaming. In a way, I stumbled into the field as an undergraduate at the University of Rochester. I had the opportunity to work in the Sleep and Neurophysiology Research Laboratory, located in the medical school a short jaunt away from the main campus. The sleep lab was tucked away in an old basement corner of the hospital, a hidden gateway to the unconscious mind.

It was there that I first learned about the electrophysiology of the sleeping brain – the waxing and waning patterns of electrical activity that occur as the brain descends through deeper and more active stages of sleep. While the science of sleep is, in its own right, a fascinating area of study, I was especially interested in dreaming, with entering and understanding that dark basement corner of our unconscious mind.

In Rochester, the bedrooms in the sleep lab were near luxurious, with large queen mattresses, thick warm duvets and art deco bedside lamps; artistic paintings of sleeping figures adorned the walls. The control room where we monitored subjects was set apart from the bedrooms, separated by a long hallway, a private bathroom, a kitchen and a lounge area. The biggest challenge for me was always staying awake all night while observing hypnotic brain waves on monitors, watching infrared videos of subjects' sleeping selves.

Into the Dream Lab

Having the control room separated from the bedrooms meant we could easily keep the lights on and talk or listen to music, while down the hall the bedrooms were near silent and pitch-black day or night – an ideal environment designed to be as conducive as possible for sleep.

In my current lab in Montreal, we have a more modest setup with two dorm-style bedrooms spotted with Ikea-like furniture. This kind of setting is in many ways familiar to our subjects, who are often students from local universities. The lab is one in a collective of sleep research set at the end of a fifth-floor hospital wing. Through a connecting doorway sits a sister sleep laboratory with three enclosed bedrooms, each with a private bathroom, where experimenters can control the lighting in the rooms to mock day-to-night phases of different lengths. Subjects sometimes spend multiple days in these time capsules, while experimenters observe how varying rhythms of light and life impact their sleep, relevant to night shift workers, blind individuals, and more.

Across the hall, the clinic diagnoses and treats patients with sleep apnoea, sleepwalking, narcolepsy and insomnia, to name a few, and our collaborations are revealing the many ways that dreaming, too, may be disrupted or treated in sleep disorders. Overall, the sleep centre is a lively setting, and the lights and sounds around sometimes leak into our subjects' dreams: from the labyrinth of fluorescent corridors on one side, to the windows outside on the other (think industrial Canadian snow removal over many nights each winter). Still, with a warm blanket, a dark room and muffled sounds from the outdoors, our subjects often drift readily into slumber.

The Scaffolds of the Dream World

Sometimes, as I sit in the control room monitoring another subject, perhaps with a fellow dream researcher, I think back to the many other labs I've worked in. During my first post-doctoral research position in the UK, for example, we had a simple lab space with two comfy bedrooms and in a makeshift manner we adapted baby monitors into a two-way communication system. The lab was set in the centre of an accessible college campus, near a beautiful coastal shoreline where subjects' dreams were sometimes startled awake by seagulls in the mornings.

In Chicago, Cambridge, and the Netherlands, the bedrooms were fashioned from pre-existing labs devoted to waking neuroscience, refitted with pull-out futons or couches to enable sleep studies. One of the bedrooms I conducted research from was tucked inside an old faraday cage, a room with a heavy metal refrigerator-looking door that isolates the subject inside an experimental chamber. Other studies in sleep clinics featured hospital-grade plastic beds and thin woven blankets, detracting from the comforting feeling so precious to sleep.

Suffice it to say, dreaming in a sleep lab can be far from perfect, and yet in my experience it has been endlessly revealing.

The most striking thing I've noticed across so many different studies, in different laboratories and countries, is that, despite the varied surroundings and comfort levels and ambient noise, there have been unexpected similarities in my subjects' dreams. Of course, while the basic sleeping environments are designed to be comparable to a regular sleeping experience at home, the unusual pressure of needing to sleep and dream well for the experiment is omnipresent. It's perhaps not surprising, then, that almost half of our subjects dream about the experiment itself.

From back in the control room, as soon as I see a subject enter REM sleep – with the characteristic rapid eye movements that define the sleep stage – I try to imagine what they might be dreaming. I wait five to ten minutes for their brain and muscle activity to ramp up, then grab my microphone and carefully, gently, call their name to ask, 'Can you tell me what was going through your mind before I called you?' I pause and wait, as they usually stir for a bit, and hmm and umm as they try to pick up the pieces of the dream that's just fallen away. In regular life, it's relatively unusual to be woken in this way, but the gentle interruption is precisely what gives us prime access to the dream's contents. I listen carefully to their halting words, as they strive to tell me everything they can remember. A dream:

> I see the wires on my face. There are hospital corridors that I hover backward through. The student researcher follows me and I want to escape her. Then I'm in the lab and my parents come into the room. I am annoyed, I only have minutes to fall asleep.

Or:

> I dreamed that the experiment was going on too long. I felt like I wasn't performing well enough on the sleep side. I remember wondering (in my dream) if you could possibly see my dreams through the electrodes and if you were going to find me weird.

The Scaffolds of the Dream World

Poring over a decade of studies set in Montreal, my team and I uncovered references to the lab and to the experiment scattered throughout hundreds of dream reports.[1] Subjects dreamed of the study setting, of the corridors of the hospital, and experimenters appearing as dream characters; they dreamed of their mission to sleep well and to remember a dream, of objects like electrodes or computers or clipboards; and finally general sleep-related themes such as wearing pyjamas or seeing a bedframe. Though several of these study elements often appeared together, dreams never re-enacted the presleep laboratory episode in full. And this is a fact of dreaming: fragments of recent experience are often woven into dreams and combined with other memories to create a novel story, but single memories are almost never replayed in their entirety. One exception to this is the special case of nightmares, where traumatic memories are re-created in part or in full, but this represents a breakdown in the normal process of dream creation – and we'll return to these cases later.

While the lab dreams we observe do vary considerably, there are several recurring features that appear across many different subjects, suggesting the presence of certain design pressures that shape the narratives of dreaming.

Dream scenes in the lab often included research personnel appearing alongside other characters, especially friends or family, who crop up as part of fictional social scenarios in the lab. In fact, there is nearly always a social element to dreaming. In home dreams, social situations occur in over 80 per cent of reports, and up to 25 per cent of characters are family members and 20 to 40 per cent are friends. In studies conducted in the lab, research personnel appeared in over half of our lab-related dreams – these are virtual strangers who the subject has only just met. This brings us to our first key element of the scaffolding of dreams: dreams are consistently social in nature, encouraging dreamers to freshly re-engage with people, even strangers, from their waking lives.

This social design provides a helpful clue about one function of dreams themselves: that dreamed social simulations may have evolved to support our development as a social species.[2] That dreams are selectively social, that is, they overrepresent social elements of our waking lives, could serve a purpose in reinforcing social skills overnight, perhaps allowing us to test out interactions and their potential consequences during sleep.

Of course, visiting the sleep lab is not your typical social affair. Sleeping in the lab is unusually intimate, with an experimenter watching as a subject sleeps and recording their dreams. It's not entirely surprising, then, that subjects also dream of being observed and of wanting to perform well for the experimenter. This ties in to a second element of how dreams are designed: dreams are often performative in nature, fashioned around a theme of skill rehearsal or achievement. Lab dreams often incorporate the basic task of trying to sleep well and remember dreams, along with other experimental tasks, too.

One subject dreams:

> My mother was with me, we walked in a corridor into the testing room. [The experimenter] put electrolytes in my head to write the dreams directly on paper. I

The Scaffolds of the Dream World

went to another corridor to do the second test ... we wanted to finish to get out of the laboratory.

Another reports:

I was trying to fall asleep and not doing very well, because there was no real barrier between the bed and the researchers ... and at one point my bed was in fact outside. The researchers referred to my brain patterns as a 'Dante pattern'.

Though there's no such thing as a Dante pattern, the latter subject's dream of trying and trying to sleep and perform well, despite many obstacles, is very common in the lab. It seems that while awake we make note of certain challenges or tasks at hand, and once asleep we are again put in a simulation to try to achieve our goal. The dream is often repetitive or circular. This subject repeatedly tries to fall asleep, but one thing after another gets in the way; that subject is doing tests for the experiment, but after finishing one test a second comes up.

In dreams collected at home, we witness similar themes of perpetually attempting (and often failing) to achieve a goal, like trying interminably to catch a bus or to get to the airport, endlessly looking for an object or outfit, underpreparing or arriving too late for an exam. These dreams seem to be goal-oriented and, like in the lab, they seem to be more about trying rather than succeeding, dreams of process rather than completion.

Beyond simply reflecting desired goals, research and perhaps common sense tell us that dreaming enables nightly rehearsal in useful skills and that this repetition may be key to learning. While asleep we can practise using our dream bodies, to manoeuvre dream objects and navigate dream worlds. For instance, athletes and musicians dream of practising their particular sport or instrument, sometimes in high-pressure dreams of performance. If our lab dreams are any indication, another skill commonly enacted by the dreaming mind is that dreams are often exploratory – we are seeking, searching, probing through an unfolding map of the world.

Consider the following:

The researcher came into the room, turned on the light, and another person came to talk to me. She cared for conscious people in dreams. The bedroom had another door with a corridor, and there were people in white uniforms at the end of this corridor in a room.

Another example:

The experimenter comes into the bedroom. I see in front of me, where there should be a wall, a door. The door is opened by Jim from *The Office*. I see on the other side of the door a small corridor, and another open door. Jim tells me I can get up.

In these examples, dreamers imagine waking up in the lab and peering out into the expanding scene. The dream seems to reinstate a world that extends out from us spatially and temporally, like ripples of space and time. Some subjects even dream of making their way home from the study, weaving through hospital corridors and traversing city streets. These dreams build on our ability to remember the recent past, including where we are and what we're doing, and project ourselves into possible near futures, a form of mental time travel with a hint of absurdity.

While some scientists quite fairly argue that dreams collected in the sleep lab might not represent natural dreams – that they're not the same as dreams collected at home – in my opinion they are still informative and useful, and they help flesh out the key pieces of scaffolding for the dream world. In fact, dreams collected from home and survey studies reflect similar themes to those collected in the sleep lab, fashioned around social simulation, skill rehearsal, exploration and reference to the current environment and ongoing concerns.

Another piece of the puzzle, so to speak, of how dreams are designed and built up comes in the form of 'typical' dream themes.[3] These are those dreams whose symbols or narratives are found repeatedly and with high prevalence across the population, and even across different cultures and ages. You have almost certainly experienced one or some of these typical dreams yourself. Upward of 70 per cent of people report experiencing the most typical dream themes like being chased, or falling, or trying again and again to do something. Over forty themes have been uncovered in survey studies, and many of them expand on the designs we've identified in lab dreams.

The social design of dreams, for instance, is evident in typical themes of being chased or pursued, bad dreams of infidelity or pleasurable erotic dreams, and the semi-social dream of sensing a presence in the room. Other typical social dreams seem to cluster around concerns of self-consciousness or embarrassment. This seems to be the common thread to dreams of being inappropriately dressed or being nude, which resemble lab dreams of being observed by experimenters who are peering into the bedroom or recording subjects' private thoughts.

Other typical dream themes are more positive and exploratory, such as finding money or – my personal favourite – discovering new rooms or passages in your home. Sometimes, this dream plays out fantastically, like opening magical doors to another world. Other times, it's much simpler, like noticing a new closet in the hallway or a useful dishwasher in the kitchen (a good dream!). This theme is similar to lab dreams where unexpected windows, doorways and corridors keep appearing as subjects wander through the hospital setting. It seems like, rather than starting with a predetermined map, the dream world is continually created wherever the dreamer ventures, unlocking new spaces along the way and encouraging further exploration.

While typical dreams are so named because of how often they occur in the general population, it seems like each of us dreams a careful selection of themes that cluster around personal concerns. One person might recall frequent social dreams of embarrassment, like being naked in public; another person might have typical dreams of failure, of attempting to achieve a goal again and again without success. These dreams can take on a recurrent nature, repeating many times over many nights, and reappearing more frequently during times of stress. Up to 75 per cent of adults experience recurrent dreams like this, which can begin at a young age and persist for the rest of one's life. To give one example, the typical dream of missing an exam often begins during school years, when academic stress is at an all-time high; but this theme can recur for years, perhaps reappearing before a big presentation at work or an interview. Although the circumstances are different, the stress of performing well seems to trigger the familiar dream scenario.

While recurrent dreams are common to a large number of people, they seem to be more pathological than typical dreams; we'll return to recurrent dreams later, because they overlap quite a lot with nightmares.

Coming back to typical dreams, these universal themes reveal certain consistencies in dream formation across many people. Like lab dreams, they reveal a dream world that seems to be constructed around several purposeful designs – pressuring our dreams to be filled with characters to interact with, tasks to accomplish, avenues to explore. These themes point towards a sort of structural foundation to dream formation and provide clues to dream function itself, as we will continue to discover. At the same time, typical dreams are curious because they reappear in so many cultures and ages, and yet are often *not* experiences typical of waking life. When have you ever discovered a new room in your house, fallen or flown through space, or found yourself nude in public? And yet, a good deal of people will experience these dreams, possibly even recurrently.

What does this tell us about the design of the dream world? It's hard to argue, for the more unusual themes that seem far removed from waking life, that they play a role in reinforcing specific skills, or even any relevant social practice. To understand a bit more about some of the more unusual and unexpected dream themes, things like teeth falling out or being unable to find a toilet, we'll have to look at another source of dream formation: the body itself.

The body's dreaming, too

It's not unusual to realise on waking that elements from our surroundings have found their way into our dreams: a barking dog, a car alarm, a chorus of birds, all transformed into the sounds and setting of the dream world as our brains try valiantly not to wake up. Needless to say, if the stimulus is too intense it will provoke an awakening, which is of course the basic principle of an alarm clock. But besides these morning interruptions, most of us consider sleep to be a time when we are cut off from the physical world, encased in our minds and insulated from our sensing bodies.

In actuality, dreaming is not so isolated from either the sleeping body or its sensory access to the world. Instead, there continues to be a flow of information through our senses, which is at times fluidly incorporated into dreams.

The influence of the body is perhaps most noticed in dreams at times when it's least desired: the urge to urinate, the inability to speak or move, grinding teeth or ringing ears invading the peaceful blanket of sleep. An extreme example of this comes in the form of sleep paralysis, where you are unable to speak or move in the moments before fully waking, often accompanied by frightening dreams.

The Scaffolds of the Dream World

I remember the first time I had sleep paralysis: early one morning before high school, I woke to the terrifying realisation that my whole body was completely immovable. I strained to open my eyes and even tried to scream but was unable to lift a muscle or make a sound. I had no idea what was happening, and in a panic I noticed a shadowy presence near the door of my bedroom, slowly coming towards me. I thought I was going to die. Then suddenly my eyes lifted open, and it was daylight in my empty room. I gasped to catch my breath, confused and afraid.

I went on to experience sleep paralysis regularly for years, and it was a primary reason I became intrigued by sleep research in college. I was reassured to learn that sleep paralysis is neither dangerous nor uncommon (up to 40 per cent of the population experiences it at least once); it's simply a relic of the natural muscle paralysis that occurs during REM sleep. While it may seem strange that the body is physically paralysed during REM sleep, this has a functional benefit: it inhibits us acting out our dreams in our real bodies. On the downside, this paralysis sometimes seeps into our dreams, especially in the liminal space between sleeping and waking, and experiencing this can range from alarming to downright terrifying.

While sleep paralysis is one extreme example, in fact, the body is often a part of dreaming; it's just that most of the time, we don't notice it in this way. In many ways, the body physically experiences what is happening in the dream, and vice versa, as the dream also responds to what occurs in the body. In other words, dreaming is not generated solely by the brain; it develops in communication with the sleeping body.

Into the Dream Lab

The first known physical correlate of dreaming was rapid eye movements, discovered in 1953 and thereafter defining the stage of REM sleep.[4] When subjects were awakened in periods containing these 'rapid, jerky, binocularly symmetrical' eye movements, 74 per cent of the time they reported vivid and visual dreams. This discovery led to a surge in dream science studies (although we've since learned that dreaming can occur in other stages of sleep), and kick-started a tradition that continues to this day, of attempting to link objective indicators – like bodily and brain activity – with the mental experience of dreaming.

Could these eye movements be revealing actual visual activity in dreams? This possibility became known as the scanning hypothesis, and some studies found support for the idea, with rapid eye movements matching visual dream descriptions, such as a recorded tennis-match dream with repetitive left-right-left-right eye movements. Though mismatches between eye movements and dreamed visuals can occur, groundbreaking studies of lucid dreaming proved that a correspondence does exist. In the late 1970s, independent researchers Stephen LaBerge and Keith Hearne showed that lucid dreamers could make deliberate left-right-left-right eye movements in their dream that resulted in matching eye movements in the sleeping body.[5] This proved that dreamers could control their physical body using movements in a dream. In the years since, many other studies have followed suit, revealing how lucid dreamers can control various parts of their physical body: they can flex their fist in a dream, which creates muscular twitches in the forearm at the same time; and lucid dreamers can control their breathing in a dream, which modulates respiration and abdominal movements in their sleeping body.

Even in non-lucid dreams, at times the content of dreams seems to physically manifest in the body. During REM sleep, for instance, the body undergoes erratic fluctuations in breathing, heart rate and muscular activity, all of which could correspond with the mental experience of dreaming. In one study, a non-lucid subject dreamed of swimming the breaststroke, and measurements taken while the subject was asleep showed that they repeatedly held and released their breath in a way that mimics going above and below the water. Nightmares, too, appear to be strongly linked to the body. In the minutes before awakening from a nightmare, most sleepers have an increased heart rate and breathing rate, as if the body is physically experiencing the stress of the dream. Of course, this can have consequences, as we'll see later, for those who repeatedly have nightmares, such as patients with post-traumatic stress disorder (PTSD), whose dreams can re-enact both mental and physical aspects of a trauma while asleep.

There is clearly some relationship between what we imagine in our dreams and how our bodies respond to or mimic our dreamed experiences. But just *how* does the dreaming mind link to the sleeping body? And what are the implications of this connection?

In the first place, we can look at the 'body map' that exists in the brain, meaning the organised mapping of sensory and motor neurons onto certain body parts. This neural map of the body seems to underlie the vivid sensorimotor imagery – the imagined sensations (sight, touch, sound, smell, taste) and bodily movements – of dreaming.

For instance, functional magnetic resonance imaging (fMRI) of lucid dreams has shown that when lucid dreamers clench their fist, activation is observed in the same sensorimotor brain regions as when they do this during wakefulness, and this also corresponds with real physical twitches in the forearm. The mind, brain and body are aligned. The same pattern is found for other types of perception like observing faces or hearing language or seeing spatial scenes – these are all associated with specific patterns of sensorimotor brain activity in waking life, and similar patterns are found during dreaming.

In general, the sensorimotor cortex, then, is providing vivid sensory and motor experience to dreams – experiences of walking, moving, flying, falling, fighting, foraging and more. It's worth noting that perceptually, dreams are most marked by visual imagery, followed by auditory and movement imagery, and less so odours and tastes; although this varies in different populations – deaf and blind subjects, for instance, have more smell and taste in their dreams, and we'll learn more about these variations later.

At times, the motor imagery of dreams is linked to real body movements, despite the intended paralysis of REM sleep. In the case of dream-enacting behaviours, sleepers physically act out dream content, often during emotionally intense dreams, such as nightmares, grief dreams, or even erotic dreams. Common dream-enacting behaviours include speaking, crying, smiling, laughing and expressing bodily fear or defensive behaviours. Fascinatingly, these behaviours can reflect the speech or actions of the dreamed self or of another dreamed figure. For instance, if you dream of flying to escape a cackling witch, you could

The Scaffolds of the Dream World

wake up shaking in fear, or you could wake up laughing, as if you were the witch. This highlights a phenomenon of dreaming that we rarely acknowledge, which is the fact that all of the characters within a dream are simultaneously created by the mind – including your own dreamed self, a friend or family member, a stranger or pet or spiritual guide. While witnessing the dream from a first-person perspective, the sleeping brain is also conjuring up these other characters with their own thoughts, feelings, actions and voices. It even, at times, uses different languages or accents for each of these characters.

This is quite impressive, that the mind is able to orchestrate the actions and emotions of all of these dreamed figures. How is it possible for the mind to do this, to coordinate such a diverse cast of dream actors?

Some scientists have looked to the mirror neuron system as one potential generator and controller of dream characters. Brain imaging studies have revealed that the mirror neuron system is active while both performing a specific action, such as gripping a mug, and observing that action in another person, such as seeing a friend grip a mug. In fact, the same brain activity can be seen when merely imagining this action in oneself or another person. In general, mirroring the actions and expressions of other people helps us to model and understand their behaviours and emotions while we are awake. In a similar manner, the mirror neuron system could also be responsible for simulating the actions and emotions of other characters while dreaming.

Some support for this idea comes from the study of mirroring behaviours – these are behavioural manifestations of mirror neuron activity, such as contagious crying or laughing (i.e., crying or laughing when seeing others do the same) and other kinds of motor mimicry like taking the same postures or facial expressions as others. Dream science studies have revealed that people who express more mirroring behaviours while awake also express more dream-enacting behaviours while asleep. As an example, people who frequently cry upon seeing others crying are also more likely to wake up from a dream crying. The same is true for smiling and laughing.

Altogether then, the sensorimotor brain and the mirror neuron system may be puppeteering the actions and emotions of our dreamed selves and other characters, in a manner similar to how we act and interact with others in waking life.

While it seems clear and perhaps even intuitive that the body might mimic or display some of the actions and emotions of our dreams, there is a potentially more interesting flip side: the physical body as an actual *source* of dream imagery.

Ever marvelled at a sleeping pet or a baby, watching as their limbs or whiskers twitch, like mini spurts of dream activity? For many years, scientists assumed that these twitches were simply a by-product of dreaming, leaking out through incomplete paralysis in REM sleep. But there is now evidence that twitches actually provide a source of dream activity. Twitches produce substantial activation in sensory areas of the brain during REM sleep. As an example, a twitch in the forearm causes strong sensory signals in the 'forearm' part of the brain-body map, as if the forearm has just experienced an intense sensation. This activation is much higher than would be expected for such a small movement; this is likely because the body is mostly paralysed in REM sleep, that these tiny twitches are in a sense amplified and can be clearly distinguished by the brain.[6] So every night when we go to sleep, these twitches allow the brain to fine-tune and recalibrate its one-to-one mapping of different body parts, almost like a call and response – twitch to forearm? Ah! there's the forearm.

This is thought to be a fundamental purpose of REM sleep, to develop and maintain our brain-to-body map over time, so that we can effectively use our bodies in the waking world. Dreaming, too, could support this function, calibrating the sensorimotor brain as we manoeuvre our dream bodies through immersive and multisensory dream worlds.

If twitches give rise to clear sensory signals in the brain, could they also give rise to strong sensations in dreams? Indeed, a common example of how dreams are influenced by twitches occurs just at sleep onset, when brief hypnic jerks are accompanied by vivid imagery. A hypnic jerk is when your legs or body suddenly twitch as you're falling asleep, and an exaggerated dream image often coincides with this sensation, such as an image of falling down a flight of stairs. The twitch seems to be magnified by the dreaming mind, perhaps because the signal is so strong compared to the stillness of the rest of the body. While speculative, twitches could also be responsible for the curious case of 'exploding head syndrome'. While it sounds dramatic, this is actually a harmless experience, where a sudden deafening noise shocks your mind as you're falling asleep. It may be that tiny twitches of the middle ear muscle (the ear's version of rapid eye movements) are suddenly amplified in the dreaming mind as the body falls into sleep. These examples show how subtle movements and sensations in the sleeping body could protrude into a dream with a potentially magnified effect.

Similar explanations have been provided for REM sleep behaviour disorder, a neurological disorder associated with loss of muscle paralysis in REM sleep. Patients often have vivid and violent dreams that are enacted in extreme and sometimes dangerous repetitive body movements, such as kicking or punching or flailing in bed. Ordinarily, the muscle paralysis that occurs during REM sleep is thought to prevent precisely this kind of dangerous acting-out of dream scenarios in the physical body. While some scientists view the violent dreams as the cause of kicking and punching behaviours, the opposite may be true: that the loss of muscle paralysis in these patients leads to excessive and repetitive twitching movements, which then feeds back into intense and aggressive dreams. In the absence of this explanation, it is hard to understand why patients suddenly develop such violent dreams at the onset of the disease.

Besides twitches and bodily movements, other internal body states could also contribute to dreaming. For instance, several typical dream themes seem related to physical sensations, such as searching for a toilet or trying to find food or water in a dream. Subjects in the lab sometimes dream of looking for food in the cafeteria or trying to use the bathroom despite unusual obstacles (such as the bathroom

The Scaffolds of the Dream World

having no doors, or personnel entering the room). Some dreams seem to incorporate the actual sensation of sleeping, such as dreams of suddenly fainting or falling asleep within a dream. And bodily states such as fever and thirst have long been associated with changes in dream quality, too, though the evidence is primarily anecdotal to date.

Even dreams of teeth falling out (a prevalent typical dream theme that almost 40 per cent of us have experienced) are associated with the experience of dental irritation in the morning, suggesting that these common teeth dreams are a result of clenching or grinding teeth during the night. This provides an explanation for why so many people have this bizarre dream theme that bears little resemblance to waking life: it's your dream's way of making sense of the ongoing sensation.

In related examples, though the texture and resistance of objects and bodies in the dream world are often inconsistent with waking reality, their very source may be real physical sensations that are incorporated into an ongoing dream. For instance, dreamers experiencing the natural muscle paralysis that accompanies REM sleep sometimes feel unable to move their body within dreams; everything feels slower than normal. Other times the body feels very light and may start to float or fly, gravity seeming almost nonexistent. Indeed, falling is one of the most commonly reported typical dreams, and flying is a seemingly universal positive dream experience. The body's position lying in bed, without any sense of vertical gravity, is quite unlike how the body feels when moving upright in waking life and may in part explain such frequent flying and falling themes (though other explanations are possible, too).[7]

In general, while real bodily sensations from paralysis to twitches continue to influence the sleeping brain and dreaming mind, these sensations are often different from normal waking life and could explain some of the more peculiar typical dream themes, like dreams of teeth falling out, being unable to find a toilet, or flying and falling.

Finally, in addition to processing bodily states, there is also continued awareness of the external environment during dreaming. Subjects in the lab report dreams of feeling the electrodes on their skin or hearing sounds from around the room. These sensations seem especially likely in dreams reported the first night of sleeping in the lab. The 'first-night' effect describes how parts of the brain stay more awake while sleeping in a new environment such as the laboratory, resulting in a less restful sleep. Some scientists refer to this as a 'night watch', where the brain continues to monitor its environs while asleep. This effect dissipates after the first night in the lab; in fact, many sleep studies discard the first night of data because of the disruption to sleep.

Regardless of how comfortable we try to make the lab, unusual sensations like wires coiled around the body and electrodes pasted to the scalp make for an objectively strange sleep experience, and these sensations often filter into our subjects' dreams. Even when sleeping at home, the sensory world often finds its way into dreams, albeit with a little less frequency. There are certain filters in place, gating mechanisms, designed to keep us from waking to habitual sensations, things like cars on the highway, talking neighbours, or the AC hum. In general, though,

The Scaffolds of the Dream World

this filtering is incomplete, and there are gaps where sensory stimuli seep into dreams.

To better investigate just how and when the dreaming mind reacts to sensory stimuli during sleep, studies have been designed specifically to probe at these gaps. Experimenters have tried presenting all sorts of tactile, visual, auditory, and scent stimuli to sleeping subjects to observe their influence on dreams. In studies set in Montreal,[8] a pressure cuff was inflated on subjects' legs during REM sleep, leading one subject to report:

> In my dream, the experimenter was there to wake me up. She turned on the lights and asked me about my dreams. I was answering her. I could feel the pressure pump on my leg. She asked me what does it feel like. I said it feels like a hug. She said, 'Doesn't it feel like someone pulling on your leg?'

In this example, the dreamer feels the pressure on their leg, and the incorporation is quite direct: the dream clearly displays the sensation and even references the ongoing experiment. Other dreamers reported sensations of tingling or discomfort in their leg, but within novel dream scenarios, like a dream of being unable to kick right while swimming. In some cases, the leg pressure penetrates dreams in a more elaborate way, at times even projecting onto another character, such as a horse trapping its leg. These cases show how sensory stimuli from the real world can be perceived while asleep and in different ways become part of an ongoing dream.

Studies of auditory and visual stimulation find similar

results, where beeping sounds and flashing red lights frequently appear within dreams. Sometimes the resulting dream matches the stimulus directly – like an auditory cue generating sound imagery. But it can also be transformed, such as a sound triggering a rather abrupt movement in a dream. It seems that the transformation of sound into movement may be quite common; in one example, a clattering noise resulted in the dream of a clown suddenly springing into a somersault.[9]

It is fascinating how stimuli can change modality (e.g. from auditory to visual movement) or even 'ownership' upon entering the dream world, projecting onto other characters and blending with the dream environment. Importantly, studies of sensory incorporation prove that real sensations are at times a source of dream content and that dreams are not only imagined sensations in the brain.

What has gone unsaid in the discussion so far is just how and why the dreaming mind accomplishes this feat, massaging and transforming perceptions into dream narratives, seemingly in real time. And what determines the broader content of dreams, where one person dreams of a horse hurting its leg while another dreams of a hug? It seems clear that sensory experience alone is not enough to construct dreams.

In the next section we will start to explore how the mind of each individual fills in the content matter of dreaming. In an illustrative case, one video of a patient with REM sleep behaviour disorder shows a man who dreams of smoking his pulse oximeter (this is a small clip on the finger that measures blood flow): The sense of pressure on the finger leads to a dream of enjoying a cigarette, in this specific individual who has fond memories of smoking. The dream emerges both from the physical sensation and a personal association from memory. A more common example can be seen in the case of flying dreams: while this sensational dream is almost universal, the specifics vary from one person to another. Where one person who lives by the sea dreams of swimming like a dolphin, another person who reads comics dreams of jetting through the sky like Iron Man. These variations are based on personal fantasies and memories associated with flying.

In fact, as early as the 1800s scientists started to establish a sensory basis for dreaming[10] and later began to unravel a formula,[11] where dreams were seen to arise from a combination of factors including sensations in the body along with recent and long-term memories, habits and concerns, too. In this view, how we perceive sensations during sleep depends not only on the stimulus itself but also on the mind and mental matters of the dreamer. Let's see whether we can disentangle some of these factors now.

The mental matter of dreams

In order to unravel how real sensations can modulate dreams, one question to consider is whether dreaming is more akin to waking perception or imagination. In other words, are dreams like real perceptual experiences of the world or are they more like thoughts or imagined experiences about the world?

To get at this question, several studies have recruited lucid dreamers who can perform experimental tasks within their dreams, designed to test whether dreams are more like perception or imagination. In the realm of visual perception, we know that in waking, humans are capable of something called 'smooth pursuit' of visual objects. If we follow a continuous moving target with our eyes, our eyes make smooth and continuous movements. But, if we try to *imagine* this movement with our eyes closed, it results in small jerky eye movements called 'saccades'. To see whether dreaming is more similar to perception or imagination, scientists asked lucid dreamers to attempt smooth pursuit of visual objects in their dream. Lucid dreamers traced an infinity shape with their thumb and followed the movement with their eyes, which were being recorded via electrodes. The resulting electrooculogram revealed a smooth-pursuit pattern of eye movements as they continuously traced the infinity shape, suggesting dreamed vision is more akin to perception.[12] In other words, the things you see in your dreams are registered in a way that is similar to how we see things in waking life rather than as figments of your imagination.

To further back up these findings, we can look at other forms of perception. It turns out that, just like vision, our perception of time is similar in dreaming and wakefulness. To investigate temporal perception in dreams, lucid dreamers were asked to indicate with eye movements the passing of ten-second blocks of time from within the dream (ten seconds, twenty seconds, thirty seconds), and their metrics were accurate to real time.[13] This goes against the commonly held belief that large chunks of time pass within a single dream and suggests that our perception of time in lucid dreaming is relatively true to waking life. More recent landmark studies of lucid

The Scaffolds of the Dream World

dreaming have shown that dreamers can even accurately understand speech – hence auditory perception – while asleep and are able to fully comprehend and respond to simple questions from within a dream (but more on that later). These studies amount to a view of dreaming that is different from what we once thought: dreamed perceptions are experienced in a similar way as real perceptions in waking life.

Nevertheless, the unique neurobiology of sleep interferes with some perceptual processes. As we go about our waking lives, we are continuously processing and combining multiple streams of sensory information. This is called 'binding', where our brains collect features of the world and integrate them into the coherent objects we perceive. This allows us to experience a comprehensible world, rather than a series of disembodied shapes, colours, motions and sizes.

In dreaming, things are a bit different. We are all familiar with confronting examples of imperfect or at least unusual bindings in our sleeping lives: your best friend appears in your dream with glowing blue hair, or the book in your hand is suddenly a cell phone. In part, this is because there is a deactivation of parts of the brain important for sensibly integrating different strands of information (the inferior parietal cortices). Because your brain has partly lost some of its ability, this can lead to inappropriate or atypical object-feature bindings. Luckily, within dreaming we tend not to question these bizarre image pairings, and this is likely due to another unique pattern of brain activation during sleep: inhibited prefrontal brain activity. The prefrontal cortex is responsible for higher cognitive functions in the brain and so would play a part in recognising when a collection of features that make up an object is, for lack of a better word, weird.

Although the combined features of dream objects can be bizarre, we nevertheless experience them as whole, as making sense within the dream, despite being unusual. So where do the features that bind into dream images come from?

In part, as we've explored, the dreaming brain draws from ongoing sensory experiences. But there's definitely more to the story. Identifying the sources of dream imagery has been the focus of much research, and has been especially studied in microdreams – those short dream snippets that slip into mind as you drift off to sleep. Almost everyone has had the experience of nodding off while sitting on a bus or watching TV, and suddenly 'coming to' as a bizarre thought or image erupts into mind. In our daily lives, and even in dream science, these little dreams are often disregarded. But carefully studying these microdreams has revealed that often these brief images are composed of three things: ongoing sensory stimuli, recent sensory impressions (a colour or shape or touch observed), and broader memory stores for context, including memories from both the immediate and more distant past.

One technique for collecting microdreams is the 'upright napping' procedure, basically allowing yourself to repeatedly nod off while sitting upright, like what happens when attempting to sleep on an airplane or on a bus. This is one technique that prominent dream scientist, and my personal mentor, Tore Nielsen, has relied on

The Scaffolds of the Dream World

to unravel the makeup of microdreams.[14] In one example, while at a conference and dozing off, Nielsen had the following microdream:

> A heavy door made of wood suddenly swings open and slams against the corner of a countertop.

Just prior to this dream, Nielsen was observing a PowerPoint slide depicting a large, closed, brown wooden door; then, at the moment of sleep onset, the conference speaker made a thudding sound by hitting the microphone. Together, the sound and image merged into the door-slamming dream. But wait. How is it possible that the dream of a slamming door occurred at exactly the same time as the unexpected microphone thud? If we look to waking perception again, we find some clues to explain this phenomenon. Our waking brains continually modify our perceptions in peculiar ways, presenting us with a seemingly more coherent experience. As an example, if you see a video of two blocks hitting each other, and a sound is played slightly before this visual collision, your brain will actually delay perception of the sound in order to match the timing of the blocks hitting. In a more common example, if the audio on a TV program is slightly delayed compared to the picture, you will nevertheless perceive the speech as occurring in sync with the actors talking. In the 'ventriloquist effect', we perceive a dummy's mouth producing a ventriloquist's voice, meaning our brain shifts the perceived location and timing of a sound to match the visual scene.

This is known as multisensory integration, where our attention binds multiple streams of sensory information in a way that makes sense; this is at times an illusion, a trick of the mind that shifts and alters our perception to create a coherent image. In the microdream example above, perception of the 'thud' was essentially paused until it could be explained by the new dream of a slamming door. In other microdream examples, a tennis ball 'whop' led to a dream of an arm being slapped, or a sudden dip in airplane turbulence coincided with a dream of a passenger spilling a glass of wine. Our sleeping brain merges real sensations with recent perceptions (e.g. seeing an airplane passenger or a door) and broader memory, too, like the prior knowledge and experience that tells us that doors make a thudding sound when slamming.

This capacity is not limited to microdreams; it is also evident in studies of lucid dreams, when audio or visual stimuli become integrated into dream scenes. For instance, when I presented flashing red lights to lucid dreamers, one of my subjects reported that in the dream, *'I could tell when the red light came on because it got hot and the sun got brighter.'* In this case, the light cues were perceived as part of the dream image of the sun. And we see another level of integration here: when the sun gets brighter, the dream also becomes hotter. This is based on broader memory, on knowing that the brightness of sunlight is associated with warmth. In another example, when a series of beeping sounds were played while a subject slept, they later reported that in their dream, *'I was shopping in a supermarket and I could hear the beeping, and it was like I was getting loads of messages on my phone telling me what to buy... things like, "buy some biscuits."'* Here, the beeping sound is perceived as coming from the phone, and it further creates actual semantic content based on prior knowledge that beeping phones usually contain messages. The brain is trying to make sense of its experience: 'Where is this light from? What is this beeping? It must be a part of this world I'm dreaming.'

In this way, our memories and knowledge add context and form to the sensory features of dream imagery. This can be seen in sleep paralysis episodes, too, where our fearful experience of paralysis is projected into dreams of a malevolent presence, a threatening figure that is somehow responsible for our immobility. Our mind tries to make sense of the situation: *'There must be some terrible reason I cannot move.'* What's more, the actual form that this shadow figure takes is unique to different cultures. In cultures that believe in witches or ghosts or aliens or devils, these are the figures that people perceive during sleep paralysis, as memories instilled by culture shape the content of their dreams. For instance, in the United States, it's thought that many accounts of alien abduction could actually stem from sleep paralysis episodes; the mind tries to make sense of what's causing its paralysis, and in a country with more UFO sightings than anywhere else in the world, alien abduction seems most likely. In other cultures, like Egypt, for example, sleepers perceive the Jinn, which is a supernatural demon from Islamic mythology. In China it's more common to perceive ghosts, whereas in Newfoundland there's the Old Hag, a travelling spirit from folklore who sits on the sleeper's chest. In all of these examples, the physical sensation of paralysis is projected into a form that makes sense, according to cultural memories and beliefs.[15]

In sum, even as we sleep our brains are constantly binding multiple streams of information in order to present us with a coherent experience – integrating sensations, together with recent memories and past experiences, and broader knowledge and expectations about the world, too. These are all combined sources of dream imagery.

As a final point, studies of dreaming also depend on one other factor, which is attention, specifically one's ability to pay attention to, recall and report one's dreams. Everything we've learned about dreaming so far we know only from subjects' reports; after all, we have no other way of recording dreams (though this may change in the future, as we'll see later). For now, to study dreams we need subjects to first encode their dream experiences in the sleeping brain, then remember them after waking, and finally report them with some level of detail, capturing the dream's emotions, sensory vividness, narrative structure, and more. All of these steps require skills of attention and memory.

From brain imaging studies we know that 'high dream recallers' – those people who recall their dreams more often than most – have greater white matter density and brain activation in areas of the brain associated with attention and memory, such as the medial prefrontal cortex. And high dream recallers also perform better on tasks requiring visuospatial attention such as the 'mirror-tracing' task, where you have to trace a figure on paper by observing its reflection in a mirror. Those who perform better on this task, and have better visuospatial skills, recall more of their dreams. Verbal fluency, a measure of how easily someone

The Scaffolds of the Dream World

can produce coherent speech, is also key to reporting dream experience, and is correlated with dream recall. In other words, your ability to pay attention to, remember and report on an experience factors into how much you can generate, recall and report dreams.

Despite these individual differences and where you fall on the spectrum, training attention and memory can reliably increase almost anyone's dream recall. This can be as simple as establishing the habit of recording dreams in the morning and setting the intention to remember dreams each night. Research has repeatedly shown that keeping a dream journal and cultivating positive attitudes towards dreaming increase dream recall.

The circumstances around an awakening also influence dream recall. Any distractions present on awakening will interfere with memory for a dream, so something as simple as remaining still with your eyes closed while recollecting a dream will increase how much you can remember. Subjects recall more dreams if immediately prompted for a report (rather than getting out of bed before reporting, for example). And while recall is highest in the morning, many people report spontaneous dream recall that seems to be triggered by events throughout the day – my dreams tend to resurface the moment I return to bed each night, as if the familiar proximity to sleep sparks the memory of a dream. Undoubtedly, many, if not most, of our dreams simply go unremembered.

This begs the question: does increasing attention to dreams, such as by keeping a dream diary, simply reveal dreams that were already occurring but usually forgotten? Or do practices for improving dream recall actually change the content of our dreams, generating more vivid and perceptual dreams as a result? To start to answer this question we can look to the case of 'white dreams', where you have the feeling of having dreamed but cannot recall any specific content. Up to one third of awakenings are associated with white dreams and many people assume these are cases where they had a full dream but forgot it by the time they woke up. But there is another possibility: white dreams could be minimal dreams that lack much perceptual detail to remember. An example would be a dream of occasional thoughts or vague images going through the mind or a sense of time passing without any further story. Waking from these dreams would leave the feeling that we were experiencing something, but without remembering much content. This resembles what's called a 'weak glimpse' in waking perception: when subjects are shown visual objects that are almost completely masked by a filter, they often report the experience of having seen an object but cannot identify or remember what it is. They know they experienced *something*, but their perception was too minimal to comprehend or recall.

Studies of waking perception can help us to interpret white dreams and minimal dreams even further. In waking life, our perceptual experiences vary dramatically in how vivid, clear and stable they are. To illustrate, if we observe a tree from a short distance on a clear day, the tree is clearly visible, solid and unmoving. But when we view a tree outside a rain-splashed window in a moving car, the tree becomes blurry and fleeting due to the overlapping images and motion. As anyone who has ever dreamed can attest, these perceptual qualities diverge dramatically in dreaming. For instance, some dreams encompass our entire visual field and are saturated with colour and brightness, whereas other dreams consist of small, faded images lacking focus, colour or luminance. Dreams can also be markedly discontinuous, with images that change fluidly over time, and dreamed perception can lack clarity: we may see something that looks like a fox or a wolf, or is it a dog or some other animal? At the lowest end of the perceptual spectrum are dream images that lack any vividness, clarity, or stability. This could be the white dream, a minimal dream that lacks much detail to recall.

We also know that minimal dreams can occur in the form of sleep misperception, which is when subjects mistakenly feel like they have been awake despite actually being asleep. Subjects sometimes have the impression that they were lying awake in bed thinking when really they were *dreaming* of lying in bed thinking. This occurs more often for people with insomnia, who have restless sleep and are more likely to feel awake even while sleeping. We'll return to these cases when we uncover how dreaming contributes to different sleep disorders and how dream interventions can help. (For instance, sleep misperception almost never occurs when subjects awaken from vivid perceptual dreams, which leave the strong impression of having slept.)

To some extent, then, it seems like learning to pay attention to dreams not only increases their recall but also enhances the richness of dream content and improves sleep perception as well (and we will see later just how relevant this is to sleep medicine).

Altogether, dreaming is a form of perceptual experience that draws on memory and sensation to deliver a coherent, if bizarre, experience of the world – one that varies in intensity, relies on attention and follows certain purposeful designs. In dreams we inhabit avatar bodies and are immersed in simulated realities: rehearsing skills and building social bonds, navigating the labyrinth of the mind. Our prior knowledge and memories fill in these scaffolds – the functional designs of dreams. Over time and experience, dreams serve as a fertile playground for developing skilful means of being in the world, and we'll learn more about how dreams function in the next chapter.

On a broader scale, one boon of modern dream science is that dreams can be collected from many different laboratories and studies, so we can collectively examine their characteristics and patterns. This has led to a consensus that dreams are novel creations, constellations of related sensory and memory bits from waking life. Dreams sample from and build around our experiences of interacting with, performing in, and exploring the waking world. And yet, dreams are also unlike waking life in consistent ways, with unusual typical themes that reoccur across cultures and eras, perhaps in part having their basis in our kindred human bodies. Over time, we each grow accustomed to our own brew of bizarre and implausible dream themes (like flying or meeting the deceased) and realise in relating to others that these unlikeliest of scenarios are in fact nearly universal.

Understanding dreams as products of our bodies and memories is the first step towards an approach for designing dreams and nightmares and using them to achieve

The Scaffolds of the Dream World

certain outcomes – the ultimate goal of a dream engineer such as myself. But we'll need to explore a little further to see how this can be accomplished.

2

The Dreaming Brain

Going into graduate school, I landed a research gig with Tore Nielsen at the Dream and Nightmare Laboratory in Montreal. Most of our research used polysomnography (PSG) to learn about how the electrical activity of the brain and body corresponds with dreaming, and we designed studies to uncover the types of memories and emotions most often experienced in dreams. In laboratory studies, one of the primary methods used to explore these links is to give participants simple learning or emotional tasks before sleeping – pared-down forms of real-world learning, like memorising vocabulary or watching stressful videos – that we can then witness sprouting into dream content.

Today, science has established that almost every waking function is actively improved by a night of sleep and that sleep is especially important for our cognitive health and emotional well-being. The question of whether *dreams* are connected to these functions, however, is still an open one. At the very least, it is clear that the content of dreams is related to memories and emotions experienced during the day and that dreams then impact our mood and cognition the following day. These findings provide clues to where dreaming intersects with sleep and how both sustain healthy minds in waking life.

The architecture of sleep

On a basic level, our brains cycle through four stages of sleep, each characterised by unique patterns of brain activity. Faster brain waves are prominent in more alert and awake states, and slower brain waves are indicative of deeper sleep or even when we feel tired or sluggish in the day. As we move from wakefulness through the stages of sleep, our brain activity gradually becomes slower and more synchronous:

- In the sleep lab, when participants first get into bed, we ask them to lie still so we can collect resting brain wave activity. When someone is in a relaxed state with their eyes closed, we can see fast brain waves in the alpha range (8–12 cycles per second), especially at the back of the brain, in the visual cortex, where closing the eyes is essentially 'turning off the lights' in the mind.
- Stage 1 is a light sleep experienced in the first seconds or minutes as we fall asleep, with a mixture of wake-like alpha waves (8–12 cycles per second) and slower theta waves (5–7 cycles per second). In stage 1 sleep there are also slow rolling eye movements and occasional muscle twitches as the body shifts into sleep.
- Stage 2 is a slightly deeper sleep with primarily theta waves and short bursts of high-frequency

activity called 'spindles' (12–16 cycles per second), along with occasional slower delta waves that start to appear (0.5–4 cycles per second).
- Stage 3 is the deepest sleep and is marked by an abundance of slow and large-amplitude delta waves (0.5–4 cycles per second) and is also referred to as slow wave sleep.

Stages 1, 2 and 3 are collectively considered non-REM (NREM) sleep. Over the course of the night, we cycle through these stages in a descending and then ascending order. From stage 1 at sleep onset, parts of the brain begin slowing down, then into stage 2 we see mostly theta and occasional delta waves, then stage 3 sleep has large delta waves throughout the brain; from this deepest stage of sleep we then shift back up to stage 2, a bit of a lighter sleep. Curiously, subjects in the sleep lab often wake up briefly at this point, at the end of their 'ascending' stage 2 period, and they might shift body positions and mumble, before falling back into the fourth stage of sleep, the more wake-like rapid eye movement (REM) sleep:

- REM sleep is a period marked by high-frequency brain activity mixed with theta waves. There are two types of REM sleep: phasic and tonic REM sleep. Phasic REM sleep is accompanied by rapid eye movements, muscle twitches and heart rate variability, whereas tonic REM sleep is relatively quiescent, and both are accompanied by the by now familiar body paralysis. Similarities between stage 1 and REM sleep, including brain activity, eye movements and muscle movements, have led some authors to theorise that stage 1 is more akin to REM than to NREM sleep,[1] but this is still controversial.

The whole progression, of descending into deep sleep and ascending back up through REM sleep takes about ninety minutes and is called an 'ultradian cycle'. Graphically it's like a wave, as the brain fluctuates to deeper and then lighter stages of sleep. At the beginning of the night, this ninety-minute cycle leans more heavily into slow rhythms, having more stage 3 sleep and less REM sleep; later in the morning, the ninety-minute cycle favours faster rhythms, having more pronounced and longer REM periods but little stage 3 sleep (almost like a rolling tide across the night).

This shift over the night occurs because the ninety-minute cycle intersects with another rhythm, the twenty-four-hour cycle – or circadian rhythm. Basically, as human beings (and similar to most biological life), we are entrained to the twenty-four-hour alternation between day and night. Diurnal animals, like humans, as opposed to nocturnal ones, are driven by our evolutionary biology to stay awake when exposed to sunlight, to eat and to act and accomplish all of our goals in the external world during this period, and then to rest and undergo the many necessary functions of sleep during the night. Everything from photoreceptors in our eyes, to how our metabolism syncs with mealtimes, works like a clock to time our bodies to this twenty-four-hour cycle. Sleep, too, follows a circadian rhythm: our brains fluctuate in levels of alertness across the day, and at night they sink into slower activity and deeper sleep. Early in the night we have more deep sleep, with larger and more abundant slow waves throughout the entire brain, and we stay in deep sleep for long periods of time, ascending only momentarily up into REM. As the night goes on, though, our sleep becomes lighter overall. We dip only slightly into deep sleep, before surfacing to REM sleep for long, pronounced periods: active, dreaming, ready to awaken into the day.

There are numerous associations between the rhythms of the sleeping brain and the qualities of dreaming. Both the frequency of recalling dreams and the content of dreams are modulated by these cycles across the night: the approximate ninety-minute ultradian rhythm, as well as the twenty-four-hour circadian rhythm.

While dreaming can be reported from any sleep stage and at any time of night, morning REM sleep has come to be known as the stage associated with the most vivid and elaborate dreams. The first clues that REM sleep was more conducive to dreaming, or at least to dream recall, came with the discovery by Eugene Aserinsky and Nathaniel Kleitman in 1953, that 74 per cent of participants recalled dreams from REM sleep, whereas only 17 per cent recalled dreams when awakened from NREM sleep.[2] These results were reinforced in 1957, when Kleitman and new student William Dement (who would later go on to be a prominent figure in dream science) again found that far more subjects recalled dreams from REM sleep (80 per cent), and almost none from NREM sleep (only 6.9 per cent).[3] These early studies established the belief that dreaming was a REM sleep phenomenon. In fact, for several years research continued to show similar findings, with high levels of REM dreaming and very little dream recall from NREM sleep. This assumption became so ingrained that REM sleep became referred to as Dreaming-sleep or D-sleep.

Initially, scientists presumed that the few dream reports obtained from NREM sleep were just memories of prior REM dreams that had occurred earlier in the night. Many believed dreaming was not physically possible in NREM sleep at all. The brain in deep sleep was thought to be incapable of producing any form of cognition; it must be unconscious. This belief was dispelled by researcher David Foulkes in 1962, who conducted a study collecting dreams from across the night and revealed that dreams could be reported even from a NREM period early in the night, prior to any REM sleep.[4] It's quite remarkable how recent, and even pervasive to this day, this belief can be, viewing most of sleep (or NREM sleep at least) as a form of unconsciousness, incapable of meaningful cognition.

Today, dreams are still more frequently recalled from REM sleep (80 per cent of the time on average), suggesting that REM sleep physiology strongly enables dreaming and the necessary cognitive capacity for dream recall. However, there has been a steady increase in the amount of dreams recalled from NREM sleep over time. In part, this has been due to changing definitions of dreaming: The original concept of dreaming was reserved for the more hallucinatory and storylike dreams we often think of and see in films, rather than simpler types of mental content that can and do occur during other sleep stages. Nowadays, when we rouse a subject from sleep, we ask broadly, 'Can you tell me what was going through your mind before I called you?' The open wording invites more frequent reports than the question used in early studies, 'What were you dreaming about?' In fact, as early as 1962, Foulkes discovered that modifying the question in this way resulted in 70 per cent of subjects recalling some content from NREM sleep. Although criticised at first, similar findings by other researchers have since confirmed that dream recall is present in a majority of subjects, from all sleep stages of the night, if sampled appropriately. Thus, it is now widely accepted that dreams, defined broadly as any form of mental content in sleep, can and do occur in any stage of sleep and are not exclusive to REM sleep.

This is quite remarkable. Although most of us have the illusion that dreams are merely fleeting things that occur in the mornings when we awaken, the reality is much more impressive, that we are thinking, dreaming, and feeling across the night. I remember participating in a sleep study where I was awakened twelve times in one night to report my dreams. It was astonishing to realise just how much I was experiencing, how far my mind was travelling as I slept for those eight hours. Of course, we forget most of these experiences come morning and are left with the misperception that dreaming is reserved for only those last few moments of sleep. In actuality, dreaming can occur across the night, though, as we'll see next, its contents change along the way. Later on we'll explore further what this means: if dreams do occur throughout the night, how could this impact our sleep and mental health, even if we don't remember them? And what does this mean for those of us with nightmares or other forms of disturbing dreams?

*

The Dreaming Brain

To return to some more of the basics: across all stages of sleep, the content of dreams is largely created by the cortex, where we store the many memories of our personal lives: lovers become dream characters, childhood homes become dream settings. Sensory cortices give rise to vivid perceptual details, especially visual and auditory imagery, though dreams can also contain smells and tastes (usually to a much lesser extent). And as we learned earlier, brain imaging studies show that the motor cortex is active during dreamed actions, similar to waking actions.

While these correlates of dream imagery seem constant across sleep stages, it is also the case that there are reliable differences in the content of dreams collected from different stages of sleep, which likely correspond to changes in the brain across each stage:

- Stage 1 dreams are brief and occur in transitory states as you fall asleep or wake up, termed the hypnagogic and hypnopompic states. If collected systematically, up to 75 per cent of stage 1 sleep periods contain some kind of imagery. While often occurring in mere seconds, stage 1 dreams can have vivid visual or visceral content. These dreams often incorporate real-world stimuli, such as an alarm clock or a meowing cat.
- Stage 2 dreams are reported in 60 to 70 per cent of awakenings. While they are often immersive and storylike, they are also consistently shorter, having fewer scenes and shorter word length than REM dream reports. They also have less perceptual and emotional content, and fewer characters, places

and actions than REM dreams, and are less bizarre, include more logical thought similar to waking, and regularly feature events from recent waking life. However, stage 2 dreams also become longer, more vivid, and more narrative over the course of the night, and by morning some are even indistinguishable from typical REM dreams.
- Stage 3 dreams are recalled around 60 per cent of the time, and reports can be very short, consisting of simple perceptual or emotional qualities without narrative. Stage 3 sleep can also include minimal forms of dreaming, such as having an awareness of being asleep without any content, or simply thinking during sleep. In the 40 per cent of awakenings where no dreams are recalled, subjects sometimes report feeling as if they haven't slept at all, or they have a gap in consciousness since falling asleep. As a researcher, it's quite fascinating to rouse a participant from a long, deep sleep and have them report feeling they haven't slept a wink. This line of research is important because of its clinical implications: The *feeling* of sleeping deeply is key to how satisfied we feel with our sleep in the morning.
- REM sleep dreams are widely understood to be the most immersive, bizarre, emotional and narrative of dreams – the most dreamlike, for lack of a better term. Given that REM is more predominant in the morning, these are the dreams people most often remember in day-to-day life. REM dreams also commonly reference recent waking life concerns,

though often through metaphorical links to more distant memories, which distinguishes them from stage 2 dreams. The sensory and motor cortices are more active in REM than in NREM sleep, giving rise to the more vivid and embodied dreams characteristic of this stage. The limbic brain – including the amygdala, which is responsible for emotions in waking life – is also more active and may contribute to the especially emotional content of REM dreams, and the occurrence of nightmares, too.

Transitions between these sleep stages are fluid, ebbing and flowing from light to deep sleep and back again. Brain activity does not abruptly change from one sleep stage to the next – it shifts gradually, including some periods where the boundaries between sleep stages begin to blur. This is actually a huge hurdle when trying to categorise (or 'score') someone's sleep. The American Academy of Sleep Medicine established specific rules for how and when to classify sleep stages, and their guidelines try to account for every messy possibility: how to score sleep when someone is in stage 2 but then starts to show signs of REM activity, like an eye movement or muscle paralysis; what to do when someone in REM sleep shows signs of wakefulness, like fifteen seconds of alpha activity in the occipital cortex. These rules were first established as early as 1968, back when polysomnographic recordings were printed in real time with ink on paper, and technicians would observe each thirty-second page as an 'epoch' to determine a subject's sleep stage. To date, we still follow nearly the same rules and stick to the thirty-second window out of tradition (though new technology and algorithms are beginning to change this).

One job of dream research is to map how dreaming varies (how emotional or bizarre dreams are, for example) across the ninety-minute sleep cycle. For instance, although on average NREM dream reports are shorter than REM dreams, we can observe that NREM dreams actually become longer, more vivid and more filled with content when collected at the end of a NREM period, closer to an oncoming REM period. In fact, the brain itself becomes more REM-like as it approaches REM sleep, with more fast-wave activity and more dreamlike dreams. Along similar lines, dream reports become more dreamlike with increasing time in REM sleep; they are described as more active, emotional, vivid, narrative and immersive, the longer someone has been in REM sleep.

On another scale, dreaming evolves over the course of a night. For instance, the number of dreams and length of dream reports increase across the night, and especially NREM dreams become longer with each successive sleep cycle. This is likely due to the circadian rhythm and the fact that sleep overall becomes lighter, with REM sleep more abundant in the morning. Other qualities that amplify across the night include perceptual details, vividness, bizarreness and emotional intensity of both NREM and REM dreams. In other words, dreams become more dreamlike as the night goes on, with more realistic, immersive and socially interactive dreams occurring in later sleep cycles and in the morning. Morning naps are especially conducive to dream research: In a simple one-hour morning nap, we found very high dream recall rates

The Dreaming Brain

for both NREM and REM sleep (89 per cent and 96 per cent, respectively), with vivid sensory content in each.

Altogether, the frequency and content of dreaming fluctuates in sync with the oscillating ninety-minute sleep cycle and amplifies across a night of sleep, peaking in frequency and intensity in REM periods and in the morning.

So far, we have limited our exploration of dreaming exclusively to the sleeping portion of our lives, considering only how dreaming varies during the night. However, advances in technology now allow us to record a much vaster array of data than previously possible. In particular, modern high-density EEGs (electroencephalograms) allow us to monitor the brain in high resolution, to fine-tune our understanding of what is happening in specific areas of the brain at specific times. In this way, dream research has shown something unexpected: dreaming processes may not be exclusive to sleep.

While traditional sleep scoring uses six electrodes placed on the frontal, central and posterior regions of the scalp, modern high-density EEG can record up to 256 different electrodes at a time. This has led to the discovery of 'local sleep', where slow delta waves – those waves that are typical of a brain in deep stage 3 sleep – appear over one region of the brain even when the rest of the brain is awake. This means that dreaming can happen not only in any sleep stage but also during waking mind-wandering episodes, or what we might more commonly recognise as daydreaming.

Instances of local sleep usually occur when slow waves start to appear over the medial prefrontal cortex. These slow waves are indicative of essentially falling asleep in parts of the brain while remaining awake in others. In this case, the part falling asleep is the part associated with metacognition, that is, the ability to direct and control our thoughts. When we are fully awake, we enjoy a certain level of metacognition, but as we let our minds wander, we start to lose this ability; our thinking becomes less focused and directed. Of course, as slow waves become even more prominent as we drift into sleep, the metacognition available in dreams decreases even further.

To illustrate this progression: as we move from waking thought to daydreams, our thoughts might shift from arranging plans for the day to reminiscing or fantasising about a romantic date. The former requires more focused attention while the latter is more imaginative, though we retain the capacity to refocus attention when desired. As we travel into sleep, in stage 2 dreams, for instance, we become more immersed in imagery, but dreams still feature some instances of 'thinking'. An example might include a dream of playing hide-and-seek, where the dreamer is immersed but also thinking about where to hide and what to do next. REM dreams have the least metacognition and display the most active and involved imagery: the dreamer does not reflect too much on their experience (e.g. a microphone materialised so I started singing; an animal chased me so I ran away). REM dreams seem to be more about 'doing' than 'thinking', about acting without reflecting much in response to the demands of the dream.[5] This temporary release from metacognition is in part what allows us to become fully absorbed into the universe of dreaming, but it's also why we get carried away into nightmares, as we'll see later.

Overall, it seems like dreaming spans a continuum from waking thought through sleeping rather than being relegated to one state or another. In this view, dreaming is like an intensified form of mind wandering,[6] and both are derived from the same neural substrates. In fact, our lab studies have shown that dream qualities like emotion, bizarreness and sensory vividness vary from waking imagery to dreaming, with daydreams being the least vivid, REM dreams the most vivid, and NREM dreams somewhere in between. These patterns provide a window into processes occurring across sleep stages and the whole twenty-four-hour cycle. But we'll have to look a bit deeper at the sources of dream content in order to understand why these variations occur.

Dreaming and memory

When I decided to go to graduate school at the University of Montreal, a not-insignificant obstacle was that all of my coursework would be in French. With naive confidence I decided to take an immersive French course in the summer prior to starting my degree. Over the weeks, I noticed that several of my dream characters began to speak in French, some brokenly, others seemingly fluently. I could observe the wheels of language slowly reinventing my social dreamscape.

Unbeknownst to me, I was living through the design of a classic study done by Canadian dream scientist Joseph De Koninck, where immersion in a French language course was gradually associated with dreaming in the new language and with improvement in that language, too.

Into the Dream Lab

In the realm of mental function, we now know that sleep is especially important for learning, for both strengthening individual memory traces and supporting long-term memory networks. It should come as no surprise, then, that dream researchers have for some time been investigating whether dreaming, too, unlocks a door to learning.

Before exploring further, first, a quick lesson in memory. It's important to know that memory can be broken up into different types. Some memories can be consciously retrieved, such as 'semantic' memory, which includes all sorts of general knowledge, like knowing the sky is blue and remembering the name of your favourite song. 'Episodic' memory refers to memory for specific events, such as remembering your sixteenth birthday party or what you had for breakfast last Monday. As you might guess, many of our episodic memories get lost over time, dissolved into more general semantic memory. To illustrate, though I can't remember exactly what I ate for breakfast last Monday (an episodic memory), I do know that typically on weekdays I eat toast with peanut butter before work (a semantic memory).

Both episodic and semantic memories, those that can be consciously retrieved, are called 'declarative' memories more broadly. This is in contrast to those memories that are more unconsciously learned and recalled, termed 'non-declarative' memories. This includes things like learning to ride a bike (i.e. muscle memory), and implicit associations or involuntary recall, like when the smell of the ocean elicits the memory of a childhood vacation. With non-declarative memory, we are not consciously aware of encoding or retrieving a memory – it seems to occur more automatically.

Without question, sleep plays a crucial role in all these types of learning, shaping how our brains store and retrieve all kinds of memories. Early sleep researchers thought that NREM and REM sleep were separately responsible for declarative and non-declarative memory storage. In support of this, NREM sleep seemed to correlate most with improvements in declarative tasks: memorising word pairs, recognising face stimuli and learning new vocabulary. REM sleep was associated more with non-declarative memory: improving performance on implicit motor tasks, maze learning and other visuospatial tasks. The case of semantic memory was a bit more ambiguous, as some research showed that REM sleep was important for improving in a second language, whereas NREM sleep was linked to vocabulary learning.

Over time, more complex learning tasks were found to depend on multiple sleep stages, and multiple memory types, too, so this model of attributing specific memory types to specific sleep stages was left behind. In its place, a 'sequential' model emerged, where memory was thought to be strengthened over consecutive NREM and REM sleep periods, with both stages reworking different parts of a given memory. This is now widely thought to be a more accurate view of how sleep reinforces learning, with learning dependent on both NREM and REM sleep together and the succession of these stages overnight.

In sequential models, NREM sleep, which occurs earlier in the sleep cycle, is thought to be important for stabilising recent memory traces, done especially early in the night when all the details of the day are fresh in mind. During NREM sleep, there are short bursts of fast activity launching through the brain, generated deep in the brain where memories are first encoded, and broadcasted out to more surface-level cortical neurons. Scientists think that a form of reactivation or replay of memory is occurring, that the brain is shuttling recent memory traces from their hippocampal origins, where they are first encoded, into distributed cortical regions for storage. In animal studies, we can see precise firing patterns when waking life memories are replayed exactly during NREM sleep (recorded one synapse at a time with tiny instruments), giving rise to the term 'memory reactivation.' These reactivated memory traces are slated for long-term storage.

But not all memories undergo this stabilisation. In fact, another core function of NREM sleep is in *erasing* irrelevant memory traces. Our brains accrue an enormous amount of new neural connections each day, presenting an increasing burden to the brain's resources that is not sustainable in the long term. One function of slow wave sleep is to prune through these numerous newly encoded memories and abolish all the weak, unimportant traces to make space for new ones. In essence, each night during slow wave sleep we undergo a system reset, refreshing synaptic availability in the brain and erasing quite a bit of our waking life minutiae in the process. Taken together, NREM sleep helps to preserve important memory traces and save them from erasure on a nightly basis.

So what do we know about the role of *dreaming* in this process?

The Dreaming Brain

To study memory in dreams, we often give subjects specific learning tasks prior to sleep and then try to detect elements of the task in subsequent dream content. One early and influential study, conducted by Harvard sleep researcher Robert Stickgold in 2000, required subjects to play hours of Tetris during the day, which led to around 60 per cent of subjects seeing those falling blocks behind closed eyes at sleep onset (even in patients with amnesia, who had no declarative recall of playing the game).[7] As one subject reported, there were jumbled images of *'tetris shapes floating around in my head.'*

The constant visual and spatial attention required for Tetris makes it a perfect task for getting incorporated into dreams, better than simple cognitive tasks like word-pair learning, for instance (or at least, more easily identifiable for dream research). A range of studies have since used similarly engaging tasks to go one step farther: to demonstrate that dreaming about a task is also linked to learning, to better memory for the task following sleep.

One of the first studies to show this effect, by the same Robert Stickgold and Erin Wamsley, had subjects navigate a virtual maze before and after a daytime nap. In the first place, the simple fact of sleep improved subjects' performance. But those who dreamed of the task improved tenfold more than others. A follow-up study extended these findings to a full night of sleep, and this time, performance on the maze was motivated by the promise of a monetary reward – known to augment the benefits of sleep on learning, since money 'tags' the memory as important (we'll delve into this more in a bit). To make use of the whole night, participants were prompted to give multiple dream reports: first before falling asleep, then several at sleep onset, a few in NREM sleep, and one in the morning.

In total, 71 per cent of participants reported dreams that contained some reference to the maze. Some participants mentally rehearsed the task while awake, spurred on by the idea that they would get money if they solved the maze. This waking mental rehearsal closely resembled the task itself, whereas at sleep onset, things start to get funky: *'I can see the maze, and I think if we could swim above it, we could see everything.'* As sleep goes on, the dreams become less like the task, but some are still clearly related. A subject in stage 2 dreamed of: *'standing in the middle of a maze, waiting for my friend to find me. She just kept going around and around, calling my name.'* And finally in morning REM sleep someone reported *'walking through [a maze] but it was like a formal maze, one of those outdoor ones made out of hedges and bushes.'*

Notably, these dreams do not resemble an exact replay of the task or even an attempt to solve the task. Even so, those who dreamed of the maze at least once improved more than those who didn't (and more than those who mentally rehearsed while awake), reducing their time and distance to get through the maze and solving it more efficiently.[8]

Similar studies set in Montreal have also found dreaming linked to better learning. In one study, subjects who dreamed of a virtual-reality flying task – dreams with visual or kinesthetic content related to flying – improved more on the task than those who didn't. The same was true for a Wii Fit balance task, where not only task-related dreams but also lab-related dreams (dreams of experimenters, the lab setting, or equipment) correlated with better performance after sleep.

In Rochester, too, we found that lab-related dreams were linked to better language learning. Already earlier studies had shown that dreams were tied to language learning: the more knowledge one has of a new language, the more that language appears in dreams, and the faster one progresses in learning a language, the sooner the language appears in dreams. In our study, we had subjects study video clips for fifty American Sign Language signs before a nap. Sign language, of course, is unique because it relies more on motor and visuospatial memory than other languages, using hand movements and facial expressions to communicate. We found that 50 per cent of our subjects had dreams about the lab, and these subjects had better recognition for signs after sleep. We think that lab-related dreams reflect memory processing of the entire presleep experience, including the learning task itself.[9]

Time and again, these studies show the effectiveness of 'offline' learning during sleep and dreaming. But it's hard to know whether dreaming is actively contributing to strengthening memories during sleep or whether dreams simply reflect this function of sleep.

To this end, some studies have asked lucid dreamers, who can control their dream content to some extent, to intentionally practise a learning task in their dream. In one such study, subjects managed to practise throwing darts in a lucid dream.[10] Some completed as many as thirty throws in a dream, while others faced many obstacles: one subject started using a mirror as a dartboard but got distracted trying to change it to a better target; in another dream, the dartboard was projected onto the head of a woman, deterring the dreamer from practising; in yet another, dream figures demanded money from the dreamer to purchase more darts. In the end, only those lucid dreamers who were successful in practising in their dream improved in darts-throwing in the morning. Those who were unsuccessful threw darts with a worse skill level than they had before sleep. In this case, the actual content of the dream seems relevant to the impact of sleep and suggests that the dream itself could be critical to learning.

Overall, there is now substantial evidence for a link between dreaming and learning, though there are several interpretations of these findings. The most conservative view is that task-related dreams are simply a result of underlying differences between participants. Certain individuals might have better cognitive skills or more motivation, for example, and so they dream more about a task and improve more as well. A second – and not mutually exclusive – interpretation is that dreaming reflects or actively contributes to learning. Lucid dreaming studies seem to provide some evidence that dream content itself can boost learning. This could function in a manner similar to waking visualisation, which is often used by athletes or musicians to prepare for and enhance their performance. Dreaming in this case would be a more immersive and embodied form of mental rehearsal, and potentially more helpful for learning.

Coming back to the sequence of sleep stages described earlier, a first step of selecting and storing recent memory traces seems to occur primarily during NREM sleep.

The Dreaming Brain

A second step, during REM sleep, seems to involve augmenting and strengthening connections *between* memory traces. One unique feature of REM sleep is its hyper-associative nature: REM sleep provides access to multitudes of memory traces in the brain, looking far and wide into our autobiography to relate each new experience to many similar things we have experienced before. 'Ah, this argument I had with my boss about filing reminds me of keeping inventory in my job as a cashier and also feels similar to when I got scolded by my teacher, and oh, the files are the colour of old October leaves, and the slamming cabinet is like a slamming door, a slamming phone, a shouting stop.'

Why would REM sleep be hyper-associative like this? Well, one reason is that we need to form links between experiences to better integrate each new experience into memory, to reveal patterns in the world that will help us going forward. This seems to be a key function of REM sleep, to *integrate* new experiences into memory.

In waking life our experiences are first encoded as specific episodes that occur in a certain place and time (yesterday, I ate a grilled cheese at the Blue Wolf café). If this memory is selected for long-term storage, as part of the transfer process it becomes distributed – the spatial and temporal context (Blue Wolf café, yesterday) separates from other features of the memory (a warm sunny day, a conversation with the barista, a productive writing session). All of these specifics become distributed and over time become harder to retrieve and piece together into a single episodic memory. Now I can still recall going to the café, but I don't remember exactly what I ate, or the conversation topic, or the focus of my work that day.

In fact, most episodes in memory become degraded like this with time. Only highly emotional events, or those we repeatedly recall, stay bound to their context and persist as episodic memories. This will become particularly relevant when we explore recurring nightmares later.

In part it seems like REM sleep is responsible for distributing pieces of memory and recombining them with other similar memories from the past. In a sense, REM sleep ties together fragments of overlapping memories. This leads to more generalised, semantic, or even implicit memory with time – basically, what's left of the overlap between several experiences. For instance, from all of the cafés I've visited, I now have a sense of what I like and what to look for when I go out into the city to work. In the long term, this generalisation of memory is adaptive; it allows us to retain vast amounts of knowledge, holding on to the gist of things and transforming stacks of individual episodes into a holistic understanding of the world.

How does REM sleep make these links? Well, associations between memories are formed by what's known as a Hebbian process: when two items in memory are active at the same time, the association between them is strengthened. Some items are very often activated together (like the words 'bread-butter-knife') and so they become closely associated. In waking life, this helps us to function more efficiently and decreases the brain's processing time: if someone asks me for bread and butter, I automatically hand them a knife. Of course, there are other related concepts that could be relevant, depending on the context (e.g. 'bread' as a term for money). In waking experiments, we know that subjects process more closely related concepts more quickly, and they take longer to process more distantly related concepts (differences in the range of milliseconds, mind you). What's interesting is that it's precisely these more distant or weak associations that are uniquely accessed and strengthened in REM sleep.

Before getting into the function of this hyper-associativity, let's look at some behavioural evidence. I mentioned that words are processed more quickly when they are strongly related – this is called 'priming': 'bread' primes 'butter,' 'hot' primes 'cold'. Typically, this priming effect is fastest for closely related words and then slows with increasing semantic distance. However, when subjects are awakened from REM sleep, they actually process more distantly related words *faster* than closely related words. This is highly unusual, and it is not the case for NREM awakenings, when subjects respond more like they do during wakefulness (faster access for closely related words). In another example, when subjects were required to find a link between three distantly related words (e.g. 'Falling', 'Actor', 'Dust'), they were better able to find the solution following awakenings from REM sleep than from NREM sleep or wake (Answer: 'Star').[11] As you can guess, this type of hyper-associativity is highly useful in creative pursuits, where we often need to think outside the box to find novel links between less obviously related concepts. It seems like REM sleep (and stage 1 sleep, but more on that later) is essential to uncovering and holding on to these distant links.

How does this happen? Well, the brain in REM sleep has high levels of activation all throughout the cortex, providing access to more (and more remote) memories, along with all of the interwoven connections between them. Dream scientists believe that the many unusual connections accessed by the brain in REM sleep can 'help explain the bizarre and hyper-associative nature of REM-sleep dreaming.'[12] Dreams often give rise to strangely associated images (my cat as a bat; the beach made of Play-Doh) and jump and transform from one scene to the next (my house turned into my office and then my old high school). But even across these seeming discontinuities, dreams seem to follow rules of semantic organisation, flowing through loosely related concepts and bringing them together. Importantly, the bizarreness of dreams seems based in meaningful associations (something about my cat reminds me of a bat). The concepts seem to be measurably interrelated in some way; the connections are just a bit more obtuse than usual.

There may even be more to the story. There is some evidence that during REM sleep the brain actually inhibits access to closely associated memory items, in essence avoiding obvious links between memories, those that are more predominant in waking life. A similar possibility is that the brain turns off Hebbian weighting altogether, treating all connections between memories as equal so there is no longer any sense of 'strong' or 'weak' associations. If this is the case, it would allow a free flow of activation throughout memory networks, in any which way without preference. These possibilities support two important functions: (1) increasing access to distantly related concepts that could be useful or insightful in the future (even if not typical of daily life) and (2) making sure that memory networks remain flexible and able to evolve over time.

The Dreaming Brain

The latter is important because the brain needs to reorganise even core memory patterns from time to time. Think about the last time you had a big life change: maybe you moved to a new city and needed to establish a whole different routine from one week to the next. Your brain, pretty efficiently, unlearns the old habits and forms *new* connections in its place. Overall, this is part of a key function of REM sleep: to integrate new information into memory, whether highly unusual or more of the same, and to update the programs running in waking life.

Dreams can offer insight into how this process is accomplished, how yesterday's walk through the park becomes merged with like-minded strolls a week ago, a month ago, a year ago, or even more.

First off, dreams very frequently incorporate memories of the prior day into their content, what Freud aptly termed the 'day residue'. Generations of research and common knowledge have since confirmed the phenomenon, with day residues present in 60 to 75 per cent of dreams. More generally, in over 80 per cent of dreams we can identify at least one of a dream's 'memory sources' – a waking life event from either the recent or remote past or even the anticipated future that can be detected in dreams.

To more fully explore *when* memories appear in dreams, scientists can ask a subject to identify a dream's memory sources and to provide an approximate date for when the memory is from. In a study in Montreal, one subject was awakened thirty-one times during the night, repeatedly in stage 1 sleep just at the precipice of dreaming.[13] The subject reported dreams at each awakening and was able to identify two or three memory sources for each fleeting image. As an example, one dream incorporated three pets from memory – a cat from recent life, a rabbit from twelve years ago, and a teddy bear from childhood. Dreams often combine memory sources like this from multiple time periods in life.

My team and I wanted to explore this phenomenon further, to see whether memory sources change across a night of sleep. In a study in Rochester, New York, we woke subjects up from each of the four sleep stages at the beginning, middle, and end of the night (twelve awakenings in total) and in the morning asked them to identify memory sources for all twelve of their dreams.[14] To give an example, one subject, who happened to be an EMT, reported in the middle of the night, *'There was a game of baseball with a bunch of members of my family, and two of us got really hurt.'* The memory source for this dream was an experience from two days prior of helping a patient who was hurt playing baseball at a family gathering. What we found was that recent memories like this (from up to a week ago) were more prevalent in early-night dreams, and relatively more distant memories appeared in dreams later in the night.

There was also some evidence that the same memory source reappeared in multiple dreams over the course of the night. For instance, that same subject later dreamed of *'walking home from a golf course with my dad. [I] got sidetracked playing baseball with some friends and kids that I knew.'* The baseball memory is still evident, but it has transformed and become less directly related to the initial event. Other research, too, has shown that how memories appear in dreams evolves across the night. Dreams early in the night are more literally related to the recent past or anticipated future, such as '*I was at work. We had orders coming in. I was cataloging,*' whereas dreams later in the night incorporate more distant memories and display more metaphorical reference to waking life (in the following example, the memory source is an exam): '*It's a big party with exams, people were getting called into a room one by one. Everyone was in modern Victorian dress.*'[15]

In general, memory sources seem to morph not only over the course of a single night but also over many nights, becoming less directly related to the original episode with time. And dreams often combine multiple memory sources together, linking recent memories with more distant ones, especially during REM sleep. This process follows a cyclical time pattern: after an initial spike in day residues, memories then reappear in our dreams after five to seven days, a phenomenon coined the 'dream-lag' effect. Say you have an argument with your spouse – this memory is likely to pop up in dreams both immediately and again after a five-to-seven-day delay. Dreams of the sleep lab also crop up first as day residues (the night after being in the lab) and then reappear in dreams again about a week after the visit to the laboratory. The initial dreams are more true to source: '*I dreamt of the laboratory bedroom. [It] was exactly the same as I saw it yesterday ... somebody was taking the electrodes off my head,*' whereas, in the week following, dreams about the lab are less direct: '*I am being admitted to a hospital because I am unable to remember my dreams. A team of doctors stands over the gurney.*'[16] The latter excerpt is layered within a less obvious dream about a hospital, the lab episode relating to more distant and generalised knowledge over time.

Overall, studies of the memory sources of dreams reveal a time course of memory integration and expand on our scientific knowledge about how sleep is processing memory. A highlight reel of day residues is gradually digested and absorbed through dreams and into long-term memory. Altogether, sleep's effects on memory are curious, complex and continuous – preserving, strengthening, integrating and generally sculpting the key moments of our waking lives.

Dreaming and emotion

A common mantra among dream scientists is that dreaming is a form of overnight therapy.

As is probably apparent by now, not *all* memories benefit from sleep. Emotional episodes, significant events, personal concerns: these are experiences important for memory that sleep is essential for safeguarding. *Emotion* is one key factor that determines whether an event is important enough to be remembered for the future. As an example, if emotion is experienced at the time of encoding, a memory will be tagged for storage. Our brains register emotion as a signifier that something is important. Other factors do this, too, like motivation and reward. When a memory is tagged in this way, it is likely to be preserved from the erasing effect of slow wave sleep and likely to be strengthened and integrated during both NREM and REM sleep.

While emotion is a helpful signal to the sleeping brain to record an event, it's also important that over time the emotion associated with a memory dissipates. Just think how burdensome it would be if simply recalling a negative memory always stirred up the same level of emotion as the initial event. Sleep provides two emotional functions, then: soothing our distress and holding on to the essential lessons learned.

My favourite study of all time used fMRI to record amygdala reactivity during an induced state of self-conscious distress.[17] The initial task was for participants to sing along to karaoke videos while wearing headphones. Later, while in the MRI scanner, experimenters induced self-conscious distress by playing back recordings of the participants' own out-of-tune singing. Their brains screamed of shame and embarrassment, with significant blood flow responses in the amygdala and accordant subjective feelings of distress. After a night of sleep, though, amygdala reactivity decreased in direct proportion to the amount of uninterrupted REM sleep. If REM sleep was severely disrupted, this benefit was completely lost.

This demonstrates that a key function of REM sleep is *forgetting* the distress associated with a memory, or at least dissipating the emotional response over time.

The sleep to forget/sleep to remember model claims that the unique neurochemistry of REM sleep is designed both to consolidate emotional memories (sleep to remember) and to decrease the level of arousal associated with these memories (sleep to forget).[18] As we've learned, there are high levels of activation in the amygdala uniquely in REM sleep, which supports the reactivation of emotional memories. However, the REM brain is also in a state of suppressed noradrenaline activity; noradrenaline is a neurotransmitter associated with arousal, so its inhibition during REM sleep means that we do not exhibit as much physiological arousal as we do in waking life. This unique combination of high amygdala activity combined with low noradrenaline means that the content of an emotional memory can be safely strengthened (and potentially re-experienced in dreams) in a low-arousal state. Subsequent recall of the memory then benefits from lessened arousal, too, a learned dissociation. This prevents an unwanted build-up of anxiety associated with emotional memories over time, but as we'll see later, this benefit is lost for those who experience nightmares.

To give some experimental examples: In one study, subjects were asked to memorise negative pictures (such as a taxicab accident or a vicious snake) while their heart rate was measured as an index of arousal.[19] The subjects then slept. After sleep, those subjects who had higher heart rate when initially memorising the objects later had better recall for these objects (emotion tagged the memories for consolidation), and their heart rate was lower during recall. Sleep stored the memories while dissipating the associated arousal. Other studies induced negative mood states in subjects just prior to a learning task – such as memorising word pairs – and sleep again improved memory but soothed the associated negative mood after a few nights. In this case, even though the task itself was not emotional, being in an aroused state tagged the memories for storage. (You can think of the implications here for patients with anxiety disorders, who are in a constant state of arousal, so an abundance of daily events are tagged as emotional and important.) Finally, when subjects are deprived of sleep, emotional memory suffers (such as having poor recall for negative pictures seen before), and amygdala reactivity stays high when re-exposed to these stimuli later. Sleep is thus as important for forgetting as it is for remembering.

Is dreaming, then, a form of overnight therapy? We know that emotions are ubiquitous in dreams, occurring in up to 95 per cent of dreams reported at home. Besides emotion, current concerns (like financial worries) are often found in REM dreams, as well as personally significant events (like an argument). Experimental studies have shown that exposing subjects to presleep stressors, such as watching aversive films or taking stressful intellectual tests, also influences dream content and amplifies dreamed emotions.

Dreams even stubbornly incorporate those concerns that we try to avoid thinking of during the day. The 'dream rebound' refers to a phenomenon where thoughts that we attempt to suppress during the day, especially just prior to sleep, spring back into dreams later that night. When we asked subjects to attempt to suppress a thought just prior to sleep over the course of one week – either an unpleasant or pleasant thought depending on their assigned condition – we found that unpleasant thoughts more often rebounded into dreams. But these dream rebounds had a positive impact: subjects who dreamed more of the thought rated it as more pleasant later on. This provides evidence that dreaming, too, is supporting sleep's emotional functions.

Calling back to the hyper-associative nature of REM sleep, one way that dreams may help is in providing creative solutions to personal problems, solutions we might not think of in waking life. Indeed, in one study it was only when dreams presented possible solutions to a presleep problem that subjects felt relieved on awakening, even if the dream itself was negative. On the other hand, dreams containing no such resolution were associated with a more negative view of the problem on waking.

If this is part of how dreaming works, it's likely a process that spans successive REM periods, and even multiple nights, as our ongoing concerns resurface and are re-examined in the fluid memory networks of the REM sleeping brain. This seems especially true for major life stresses, such as going through a divorce. Rosalind Cartwright's seminal work in the 1980s and 90s showed that divorced women who initially had more negative dreams involving their ex were less depressed one year later than women who did not experience stressful divorce dreams.[20] She suggested that emotional memory is being reworked over many nights of sleep. The divorce calls for a major reorganisation of memory networks, adapting to the negative experience and rewiring the brain over time. Cartwright was an early pioneer (and one of the only women) in sleep medicine and dream science in the 60s, and played a prominent role in uncovering the role of sleep and dreaming in our emotional lives.[21]

Indeed it seems like both REM sleep and dreaming play a pivotal role in maintaining our emotional health: integrating salient experiences into memory and dampening the distress associated with these memories over time. We'll explore these concepts further when we get to nightmares, where these benefits begin to unravel.

All in all, the cycles of brain activity we surf through each night have evolved for our adaptation and survival: to support our brain's immense capacity for encoding information, to recover from the wealth of stimuli and stress accrued each day, and to prepare us to re-enter the world each morning. Without question, sleep is instrumental for learning. Memory traces tagged as important are reactivated in the brain and strengthened during sleep. Those events of greatest concern and strongest emotion are most likely to be saved, becoming pieces of our personal story and teaching us how to engage with and what to expect from the world each day.

In dreams, we may be witnessing this process from within, as if accessing memory were a real experience (a re-experience). Though early views considered dreams to be random electrophysiological noise produced by the brain during sleep, today's science is revealing that dreams offer a window into sleep's mental functions, revealing which memories are undergoing consolidation and when stressful events are being processed. Dreaming is correlated with learning and with improved performance on various tasks. Some dreams unveil creative solutions to problems, a product of the hyper-associative REM state. (In fact, you can borrow from this state in the short period after waking from a dream – try using this time to reflect on a problem, for example.)

Over many nights and many dreams we see an interweaving of close and distant, recent and remote memories. This resembles a sorting process, where each new experience is filed relative to similar experiences from the past, and the dream spirals out to link more remote memories into its web overnight. With this nightly process, our sense of self evolves and updates over time and throughout our lives.

Part II

Why Dreams and Nightmares Matter

3

Why Dream at All?

We've explored some of the science around what dreams are made of and how dreams work, but the question remains: why do we need to *experience* dreaming? Surely the brain could simply organise memory and regulate emotion unconsciously while we sleep. What does our conscious encounter with dreaming add to this picture, if anything?

The answer may lie in the role of 'feelings' as a key to both generating dreams and driving dream function.

Neuroscientist Antonio Damasio describes consciousness itself as 'the feeling of what happens.'[1] It's the subjective side of our biological existence, an inner experience of the world. As a sensing being, I witness colour, sound, shape, movement. As a feeling being, I form a relationship to these sensations: what is this colour, this sound, this movement, to me? Feelings add value to the perceptions or images that evoke them; they offer an interpretation. Through feelings we come to find meaning in the world. Feelings also give rise to salience ('Is this important?') and valence ('Is this good or bad?'). These provide us with an impulse, guiding us towards where and how to act in the world.

Feelings underlie our sense of meaning not only in the external world but also in our imagination of inner worlds. In sleep, our feelings continue. We consciously feel what it is like to experience dreaming. Some feelings become so intense that the dreamer is stirred out of sleep laughing or crying, overwhelmed by imagination. If the dream world is a reflection of our everyday experience, why is that inner world sometimes amplified, in vivid – and at times marvelous or overwhelming – dreams and nightmares? Why do we revisit our very worst feelings, with traumas re-enacted and fears dramatised? From where do we conjure the nightmare or her more gracious lunar twins, all manner of lucid and euphoric dreams?

Feelings and a natural dream process

Many scientists presume dreaming is closely tied to our processing of emotion because dreaming is so quintessentially emotional and so often calls back to emotional memories.

A broader perspective frames dreaming as based in *feeling*, which encompasses emotion and also includes all manner of bodily sensation. The perception of subjective feeling relies on a sort of mind-body association, where our physical sensations are linked to memory – the smell of a perfume reminds me of my mother and brings with its scent a pleasant feeling of nostalgia, warmth and family. We fuse with each sensation a plethora of learned meaning, and perception becomes woven with memory and experience. We come to understand the outer world through its feeling – what each percept feels like and means to *me*.

Why Dream at All?

Many people think of feelings as containing only emotion. Of course, emotion is one salient type of feeling characterised by clear visceral sensations: think fear, joy, anger, sadness. Emotions are often instinctual – fear, for instance, is triggered by threats to survival – and the accompanying sensations prepare the body physiologically to respond, to fight or flee. But emotions can also be shaped through the lens of personal history. A learned fear of abandonment sees a partner being distant as a threat to survival: a fight, flight, freeze response erupts at the thought of losing someone.

Most of the time, our feelings are rather layered. Exposed to the waking world, multiple feelings rise and fall in response to the changing environment. Our mind is scanning, selecting, and categorising stimuli. Sensations in our periphery (light, motion, temperature) compete with those at the level of autonomic (heart rate, breath rate) and metabolic systems (hunger, thirst), and even ongoing emotional and psychological concerns (a recent embarrassment or relationship trouble). All of this is mixing in with the demands of any ongoing task, keeping our experience somewhat tethered and pulled towards a stable goal.

From this abundance of information, our feelings highlight constellations of meaning. We perceive elements that are salient, of relevance to our being. We do not perceive the world simply as it is. Our experience is much more dreamlike, an overlay of thoughts, images, sensations, emotions – this barrage of information that we navigate by feeling. Feeling acts as a sort of internal compass, a 'narrative line upon disparate images' that guides us through the 'shifting phantasmagoria which is our actual experience'.[2] We move towards self-relevant stimuli on the path to personal goals.

When we cross the threshold into dreaming, our internal compass remains. We continue to sense an internal perceptual surface – hunger, thirst, pain; we witness the replay of recent sense impressions, the falling snow, the drumming highway. We surf the waves of current concerns and sift through memories, relatively untethered – and absent the constraints of the external world, there is much less of a constant to anchor us.

Many dream scientists[3] describe a process where affect, feeling, or a felt sense actually guides the river of consciousness[4] – what we dream up and experience from moment to moment while asleep. They describe how reverberations of feeling through memory give rise to metaphoric dream images. Subtle and nuanced feelings, like the empty ache of loneliness, can be captured in an image of falling to the depths of an echoing well.[5] Dream images in this way can embody many aspects of feeling: emotion, bodily senses, even abstract cognitive sensations (like the curious cognitive feeling of knowing that you've forgotten something, or knowing when you've rightly remembered it).[6]

How does dreaming do this? How do we, through dreams, craft images to express such diverse feelings?

In the first place, feelings provide an undercurrent to the connections made in memory, drawing together elements through feelings of similarity. Take state-dependent memory, for instance, where entering into a certain feeling lifts familiar memories to the surface. To give a concrete example, in a study of scuba divers, when information was learned underwater, later memory was better if the divers were again underwater (as opposed to on land), or even if they imagined being underwater. In another example, the feeling of getting into bed or drifting off to sleep can tickle the recall of a forgotten dream from the prior night. Feelings are imbued beneath a memory's content, connecting the dots to other like-minded experiences.

This process, where feelings drive associations, permeates dreaming, where multiple memories and sensations are woven into images that convey specific feelings.

To illustrate this with a curious example: we know that dream characters are often inaccurately represented in dreams. Instead, dream versions of friends or family members can differ radically in appearance and behaviour from their waking life counterparts, and even shift in form from scene to scene: a good friend appears as a cat, a sibling morphs into the persona of a celebrity. And yet, a sure feeling allows us to recognise these dream characters, composites of faces from memory. In fact, dreamers rely on what's called a 'feeling of knowing' to recognise people and places in their dreams, even when these images bear no resemblance to waking life.[7] We are able to feel the undercurrent of meaning in these bizarre amalgamations of imagery.

In waking cognition, too, we experience similar 'feelings of knowing' as we sense our way through connections in memory. This phenomenon can be observed in creativity tasks, where subjects are often able to sense that they are approaching a solution in memory just prior to realising it. For instance, subjects report a 'feeling of knowing' just before solving a word associations task, sensing the link between seemingly unrelated words, sometimes feeling this knowledge on the tip of their tongue just prior to discovery. We are able to consciously experience this associative process of the mind, feeling the connections in memory. Another example can be seen in what's called 'mind pops', where a jingle or word suddenly pops into mind seemingly out of nowhere; these little mind pops are primed by associations – a simple sensation or thought or feeling encountered earlier triggers their burst into mind.[8]

This process of experiencing associations in memory permeates both dreaming and waking cognition. Of course, several neurocognitive features of sleep discussed earlier differentiate dreaming from waking cognition. One unique feature of dreaming is the breadth of memory access available, and the lack of external sensory constraints on the dream world (in other words, unlike waking, there is no constant form to the dream world, it is constructed endogenously from moment to moment). Absent the constants and demands that tether waking life, we drift freely through memory networks and enter without limits into our personal past, reliving experience through dreams.

But what is the purpose of all this feeling? And all of this re-experiencing through dreaming?

To start with, we know that in waking life, our conscious experience of feelings is functional in part because feelings are hedonically valenced – that is, they are immediately either pleasant or unpleasant. This valence carries implicit meaning and directs our actions in the world, either to seek out and improve upon pleasant states or to eliminate and avoid unpleasant states. Even the most basic feeling of thirst or coldness or fatigue makes demands on us to do something: we drink to quench thirst, sleep to quell fatigue. Feelings thus have an inherent purpose: they are signals of need that arise in consciousness to prompt a necessary response.[9]

This is a primary function of feeling. Our desire to feel good (and not bad) entrains us to fulfill our bodies' needs. Feelings accomplish this by being nuanced and precise: the badness of anger is different from fatigue is different from loneliness. These feelings differ in personal ways, informed by history. And through experience, we learn how to reconcile these feelings via interactions with our bodies, with other people, and with the world around. Even more, we learn through imagination and dreaming, too.

To first look at waking imagination, consider that at any given moment, there are countless feelings we are carrying: emotional and bodily needs compete as we move about and respond to the demands of living. Concerns about relationships or work are hovering on the mind, and our imagination is often grinding through spurious scenarios, gauging how to manage these situations. We imagine what it would be like to react with this or that behaviour or comment: how it would feel, what would happen next.

Dreaming displays a similar process of exploring and enacting imagined scenarios, albeit in more dramatic fashion than waking daydreams. Dream feelings can act as a sort of barometer in this context: Our emotional response tells us whether the possibilities tested are important or useful to remember, whether the outcome feels optimal or satisfactory.[10] On the contrary, certain dream scenarios may present abject failures, not to be attempted in waking life. Dream scenes that induce intense feelings of fear or disgust, like nightmares, would be marked to avoid at all costs in the future.

The unusual contents of dream scenarios, compared to waking daydreams, are spurred by the breadth of memory access described already. Despite the bizarre landscape of dreaming, our feelings remain familiar and drive us to act, as they do in waking life. Perhaps more freely than in waking, we can explore the felt consequences of this or that action and seek solutions that feel satisfying. Indeed, feelings don't go away unless resolved; they linger and reappear stubbornly. (Take the typical dream of trying and trying to find a bathroom – unable to satisfy a real urinary need, we just keep searching – a classic dream.) 'Loop dreams' can also occur, where the mind plays out a particular scenario to one possible conclusion, then restarts to try another version, just to see how it feels.

Importantly, if feelings are functional they can impact the practical outcomes of sleep. In other words, this experiential nature of *feeling* dreams allows dreaming to have a function, above and beyond a simple offline function of sleep. We can first look at the implications of this claim on a practical level.

We know that during sleep, memories are reactivated, and it seems that during dreaming we experience these memories. This is not a passive one-way process, where we are simply receptive to or witnessing dreams (though several scientists have made such claims).[11] Instead, a view of 'co-creative' dreaming describes how a dreamer plays a role in determining dream content and selecting what comes next, just as we do in waking life.[12] This view runs counter to the common assumption that dreaming is something that happens to us unconsciously, that we have no control over. Even in completely non-lucid dreams, where we lack any reflective awareness or intentional control, we are still feeling and interacting with the dream consciously – we are thrust into a situation and reacting with emotions, thoughts and actions, even if absent-mindedly. This is consequential, as the feeling dreamer plays a role in producing the dream as a whole.

In research terms, we say that dreamers have some level of *agency*, that they consciously think and react and interact within dreams. In this way, our feeling, dreaming self is to some extent guiding the dream, determining what comes next.

To return to a function of sleep, recall that memories are reactivated offline during sleep, and this serves the purpose of strengthening them. Of course, to some extent patterns of memory reactivation are likely automated and systematic in the offline sleeping brain, not reliant on dreaming. But consider that this offline process is partnered with a conscious process of dreaming. There is a strong argument to be made that how we think and feel and act in dreams directly affects how the underlying memories are organised.

By directing dream content, a dreamer brings forth new activations in the sleeping brain, new memories and sensations and feelings, and shapes the storyline the dream will follow. As an example, if the dreamer shifts their attention, it will alter the visual focus of dream content; as they move to navigate the dream world, this leads to new scenery; and in reacting to other characters, emotions arise in both the dreamer and other dream figures. A back-and-forth emerges between the dreamer and the dream world.

In waking life, this is part of how feelings function. They direct us, they communicate to us; they tell us where to look, what to think, and how to behave in the world. They do this with purpose, to curate our experience from moment to moment and guide us towards our goals. Feelings similarly animate dreams with purpose: we feel an inner drive to manoeuvre dream bodies, to navigate dream worlds, and to interact meaningfully with other dream characters. Feelings allow us to experience and engage with these simulations in purposeful ways. These impressionable experiences train us for what it is like to live in the world, and inform our understanding of the world, too, shaping the very memories we are re-experiencing in dreams.

On a fundamental level, the conscious dreamer is in part responsible for selecting which memories are active and plastic in the sleeping brain, open to consolidation or reform. We know this is the case in waking life: any time a memory is retrieved or re-experienced, it becomes labile and prone to modification. We can thus change memories through re-experiencing them and, through dreaming, influence these supposed offline functions of sleep.

To take this a step farther, we can also reinforce or revise how we feel *in response* to dreamed memories. Recall that another function of sleep is to regulate emotion and dampen the arousal linked to memories. In waking life, it is another role of feelings to guide us towards self-regulation, and it is likely so within the dream world, too.

Why Dream at All?

To elaborate, in waking life, feelings are key to emotion regulation – they highlight any sources of distress and point us towards how to extinguish them. Feelings orient us on the quest to return to equilibrium, or even to a place of contentment. In the 'process model' of emotion regulation, we learn to regulate feelings using our attention, thoughts, and behaviours – focusing, evaluating and physically responding to a situation.[13] To give practical examples, in a stressful scenario, one might shift their attention to something calming or distracting, or mentally reframe a difficult problem as a useful challenge, or even actively change locations to get out of a stressful situation.

Turning to sleep, although we cannot physically change our location from the bed or turn our attention to the external world, within the dream world we are able to do these things. We can manage our feelings by shifting our attention within a dream, or by responding behaviourally, such as by interacting with other characters or confronting or running from a threat. Sometimes these strategies are successful in regulating our emotions (both in dreaming and in waking life), and sometimes they are not, which we'll see clearly in the case of nightmares.

In this view, dreaming is not a mere by-product, a side effect, of offline sleep processes; dreaming is functional in the same ways that feeling is functional in waking life. In the first place, feelings breathe meaning and purpose into the scaffolds of dreaming: We can feel our dream bodies as we move through multisensory dream worlds, reliving experiences and learning through this reliving. Dream feelings also guide us towards self-regulation within this inner world, directing our attention and our behaviours towards where and how to act and interact with the dream world. Dreaming is thus consequential, affecting how memories are accessed and how we feel in response to them, influencing these presumed functions of sleep.

We return once more to the question: Why do we need to *experience* dreaming?

I would say that dreaming adds to sleep what consciousness adds to waking life. If feeling in waking life has an inherent value and purpose, the implication is that feeling in dreams also has an inherent value and purpose. Feeling adds meaning to the information processed during sleep. It adds value beyond the biological function of sleep. Our feelings prompt us in a 'life-forward'[14] direction, attempting to guide us towards self-regulation in this unpredictable dream world, a mosaic of emotions, sensations, memories and images. How we feel and respond within dreaming plays a part in selecting which memories and emotions are active in the sleeping brain and how we relate to these memories, too.

Could all this conscious experience, this feeling and dreaming throughout the night, really be so impactful, shaping our memories and emotions before we wake?

Scientists opposed to this notion have argued that dreaming largely does not get encoded or stored in memory – we forget most of our dreams, so how can they have any lasting impact? But recent theorists suggest that forgetting dreams is in part what allows their vast function to be accomplished.[15] Think about what a wealth of an opportunity it is for learning, this full third of our lives spent asleep, where the mind can travel far and wide in memory, re-experiencing through dreams. Despite forgetting

Why Dream at All?

dreams explicitly (meaning we do not consciously recall them, for the most part), we can, through experience, retain the implicit lessons learned. And *because* we forget most of our dreams, we can keep the waking world more prominent in mind and memory, not confusing our dream life for reality.

Nevertheless, it's worth noting that many scientists still argue that dreaming is inconsequential to the functions of sleep or that dreams simply reflect the more automated functions of sleep. In my view, while we may not consciously experience every memory that is reactivated in the brain during sleep, it is likely that that which we do feel in dreams shapes the functions of sleep, organising these malleable memories and in part determining when and how our emotions are regulated in response.

Along similar lines, while much of the experimental work on dreams and learning has focused on a single task or a single night of sleep, a broader view sees dreaming as a cumulative mechanism for learning over time. Dreams allow us to explore new behaviours, to test out and revise how we feel and respond to the world in the safe and forgotten sanctuary of sleep, in a manner similar to play.[16] We learn these patterns over a lifetime of experience in both the waking world and in dreams. In other words, it's not one specific dream that teaches us, for example, the precise layout of the lab or the attitudes of research personnel; it's more of a composite of dream simulations that trains essential skills like exploration and social interaction over a lifetime of experiencing.

In a way, our conscious encounter with dreaming adds a degree of freedom to the offline processing of sleep. Even in waking life, consciousness adds a degree of freedom to our behaviour:[17] we can explore and behave in creative ways in pursuit of desired feelings, beyond mere biological need, through activities like listening to music, exploring caves, or skydiving. We seek new experiences purely for the sake of feeling. This extends to dreaming, where we can experience even non-ordinary states of being: the bliss of flying, encounters with otherworldly creatures. Through dreaming we craft new experiences, new memories and feelings, which implicitly become a part of us, for better or for worse, as we'll soon see.

In sum, while asleep we continue to experience worlds made of our minds, brains and bodies, active and alive throughout the night. We re-create simulations based in memory and sensation, and we consciously feel and experience these simulations as real. This near-constant activity of the sleeping mind has consequences, acting as a hidden variable in the functional outcomes of sleep.

This becomes all the more evident and important when we look to nightmares, as we'll see next.

Dreams, interrupted: The haunted brain

If feeling is central to dreaming, what happens when feelings become too extreme? How can this change the content, and possibly the function, of our dreams?

In some dreams, we seem to tumble around loosely associated images without much feeling. Other times, we become completely absorbed in fantastical storylines replete with blissful feelings or trapped in singularly terrifying images.

Why Dream at All?

In some ways, we can think of the *intensity* of feeling as a sort of volume control that modulates the sensory richness, the depth and vividness of dreaming. And we can think of the *valence* of feeling – positive or negative – as adjusting dream resolution, guiding the breadth of memory access and narrative elaboration in dreams. In this chapter, we'll explore in detail what happens when bad feelings become exaggerated in the context of nightmares.

If feelings usually lead us towards self-regulation, what's going wrong in the case of nightmares? Why are some of us forced to relive again and again the full force of our deepest fears, sorrows and furies, in the form of recurring nightmares? Is this a normal and healthy response to stress, or does it reveal something more problematic? And when nightmares repeatedly tear through the peace of sleep, does this irreversibly damage sleep function?

To begin, nightmares are, put lightly, extremely negative dreams; they feel more intense than run-of-the-mill bad dreams, so much so that they often jerk you into waking (whereas bad dreams are less intense and do not cause awakenings). Nightmares are not only emotionally intense but also sensorially intense, featuring vivid audio-visual imagery and even sensations of olfaction, pain, or discomfort, which are less common in typical dreams. The body itself is in a panic on waking from a nightmare: sweating, heart racing, breathing accelerated, maybe teeth grinding or limbs twitching. The stress of awakening from a nightmare is wholly unpleasant, and in the abyss of darkness, with no distractions or companions, the mind becomes a fearful enemy.

What's more, when in a sleep-deprived state, the brain's defences are weakened and the mind is unable to keep a lid on emotions, unable to keep thoughts from spiraling. The bad feelings encountered within nightmares seep into waking, disrupting one's morning or day, and even ruining sleep the following night. Because of this, many nightmare sufferers actively avoid sleep, afraid of the night and distrusting the darkness in their own mind. The nightmare seems beyond control, an attack by the subconscious in our most vulnerable state.

Why do we suffer this self-attack? Is this some kind of mental autoimmune reaction, as if a foreign object is in the unconscious that the mind is trying to eject?

In a way, the answer is yes. Researchers have long known that being exposed to traumatic and adverse events is a strong precursor for nightmares. We have shown that childhood adversity even at very early ages (before six years old) is associated with having more nightmares in adulthood. In other words, nightmares are a lasting symptom of trauma and adversity. (It's worth noting that definitions of trauma have expanded in recent years, beyond threats to survival, to include any highly stressful experiences beyond our ability to self-regulate.) As just one example, we have found that people who are deaf have more nightmares in adulthood, and that this is related to early experiences of adversity, which are especially common in deaf children, often stemming from language and communication barriers with parents and family.

There is even some suggestion that nightmare themes form early in development – before what's called the 'infantile amnesia period.'[18] Normally we forget childhood

Why Dream at All?

memories for events that occurred prior to the age of three and a half, but exposure to trauma or elevated stress in this time period has lasting consequences on memory and emotion. Those who have nightmares have better memory for their early childhood experiences and dreams. It's thought that nightmares can develop in response to early trauma and become lifelong remnants of childhood adversity. Reaccessing and reliving early trauma in the imaginative space of the sleeping mind could reinforce those same memories and feelings again and again over time.

Besides early traumatic events, in general, accumulated stress in life, such as experiencing neglect or separation in adolescence, can also trigger recurring and persistent nightmares. Even in so-called idiopathic nightmares, which occur independent of diagnosed trauma or other psychological disorders, the root cause seems to be some form of adversity.

In general, over 90 per cent of nightmare sufferers describe some of their nightmares as having repetitive themes, and these themes can often be traced back to emotional memories, to which the themes are more or less symbolically related. When asked to describe what prompted their earliest nightmares, many subjects can identify an event that occurred just prior to the onset of a particular theme, though the theme continues to recur well after the event has transpired. As an example, two of my research subjects traced their nightmares back to their parents' separation: one developed recurring dreams of being cheated on by a partner; the other started to have anxiety dreams of falling into emptiness. In both cases, the feelings expressed in the nightmares felt similar to their experiences of their parents' separation, though one example is clearly more symbolic and metaphoric than the other.

In brief, being exposed to trauma and adversity, even in childhood, leads to more frequent and distressing nightmares well into adulthood, with recurring themes that stem from their waking life memory source.

You may wonder why the brain might do this and, indeed, part of the confusion for someone suffering nightmares is not knowing why this keeps happening, why they need to keep reliving a trauma or deep-rooted fear. Nightmares seem to force into mind the very thoughts we most want to avoid in waking life; this feeling of being betrayed by one's own mind can make nightmares even more unbearable. But despite their unpleasantness, nightmares seem to reflect attempts by the sleeping brain to complete its function, to process underlying emotions, and to adapt to adverse experiences.

In one manner, frequent and recurring – or chronic – nightmares can be framed as a disorder of emotion regulation. We can see a breakdown in the usual role dreams play in healthy processing of emotional experiences, particularly as nightmares are felt so intensely that they often tear us out of sleep before the job is done. Nightmares showcase how extreme bad feelings within dreaming become overwhelming and beyond our capacity to manage or control.

This disruption can be understood first by looking at the brain components of emotion regulation, during both wakefulness and sleep.

Why Dream at All?

In the brain, regions of the prefrontal cortex are thought to control levels of amygdala activation, which dictates how intensely we feel emotional experiences. When someone is repeatedly exposed to adversity and stress, the communication between the prefrontal cortex and the amygdala deteriorates. The frontal brain regions become less effective at regulating emotion and, in turn, the intensity of emotional reactions induced by stimuli increases. In practice, this means that the world becomes more salient; for instance, a relatively harmless situation feels more threatening. Overall, the brain becomes more sensitive and does not gauge or temper perceived threats as effectively.

Disturbed REM sleep can exacerbate this process. A primary function of REM sleep is in refreshing the brain circuitry between the prefrontal cortex and amygdala each night. When REM sleep is disrupted, this process of dampening arousal cannot take place overnight. Individuals who have frequent nightmares, by definition, have disrupted REM sleep due to nightmare awakenings. They are also more physically aroused in their sleep, both during a nightmare and in its absence. For example, in the last few minutes before waking from a nightmare, many sleepers display increased heart rate, more eye movements, and quicker breathing. Even in sleep periods not marked by nightmares, chronic nightmare sufferers have more wake-like brain activity and more nighttime awakenings than those who don't have nightmares. This arousal both interferes with REM sleep function and leads to a build-up of emotional stress in the mind and body the next day. This creates a vicious cycle, as the build-up of stress in waking life then triggers further poor sleep and nightmares thereafter.

From this view, chronic nightmares are associated not only with arousal during sleep but also with repetitive sleep loss and increased emotional reactivity in waking life.

In Montreal, we conducted brain imaging research to support these claims. We collected data from nightmare sufferers as they completed an emotional task while awake, simply viewing a series of forty-eight negative pictures during eight minutes.[19] The pictures were intended to induce a state of negative arousal, with images of war-torn bodies, starving animals and natural disasters, to name a few. With two minutes remaining in the task – the period when subjects were most aroused – a radiotracer was injected into the bloodstream. Following this, subjects lay in a SPECT scanner (single-photon emission computed tomography scanner) for thirty-five minutes, which measured how the radiotracer dispersed according to where blood flowed in the brain. Subjects with severe nightmares had less blood flow in frontal brain regions, suggesting there was less activation in these areas when their emotions were running high. In other words, they seemed to be less adept at regulating their negative emotions during this stressful task.

We found similar results in a study in Wales using near-infrared spectroscopy – a method of brain imaging where infrared light is emitted over the scalp and penetrates about two centimeters into the cortex. Changes in blood oxygen levels in the cortex can be detected based on how the light scatters (when blood has more oxygen, certain wavelengths of light are more absorbed). In our study, eight light sensors were placed across the frontal cortex. This revealed that subjects with worse nightmares had less activation in frontal brain regions during emotional arousal. These same subjects also reported *feeling* more aroused than healthy subjects when exposed to negative images (as we asked them to rate their emotions during the task). Taken together, this suggests that nightmare sufferers have less self-regulation and more intense negative feelings during a stressful task.

Thus, at both a brain and a subjective level, nightmare sufferers seem less able to control their emotional response to negative stimuli. In general, this means that nightmare sufferers are more reactive and perceive stressful situations as more salient and more threatening.

In addition to causing physical arousal (a stress response in the body), emotional reactivity also dictates how we think about and behaviourally manage situations. Normally, when exposed to different stimuli in the environment, we qualify and categorise these stimuli: 'This tiger seems threatening', 'this cat seems friendly'. This is called 'cognitive appraisal', an evaluation or judgement of whether and to what extent a stimulus is good or bad, pleasant or unpleasant, appealing or aversive. Determining the threat level of different aspects of the world allows us to respond appropriately. When exposed to a friendly cat, we can approach, vocally engage, smile, and so on. When exposed to a tiger, we should respond wildly different, likely entering the fight, flight, or freeze response. These more extreme reactions are costly: we expend a lot of emotional and physical energy when under a threat to our survival.

In general, how we respond to affective stimuli can be adaptive or maladaptive. When it comes to the emotions of everyday life, we are continually confronted by little feelings of annoyance, obstacles, worries, or sometimes more severe threats to our existence. Self-regulation is an unending process where we manage all these feelings using our minds, bodies and the world around us. In order to regulate feelings as they arise, we need to find ways to resolve the underlying need. Looking for appropriate solutions, enlisting the help of others, and even seeking distractions can all be successful strategies for relegating different feelings.

However, when arousal is elevated, it signals a more significant problem and hijacks our attention and focus, prioritising one problem above others. We confer more cognitive and behavioural resources to deal with a highly arousing stimulus, to get our emotions in check, to get our body back into homeostasis. As you can imagine, it becomes increasingly difficult to regulate feelings when emotional reactivity is high and threats seemingly sprout up left, right, and centre, like a game of emotional Whac-A-Mole. And unfortunately, especially when overwhelmed, some of our strategies to attempt to manage feelings actually just make things worse.

In the case of nightmare sufferers, from the studies described above, we know that the self-regulation areas of the brain are less active when faced with negative stimuli. Moreover, there is evidence to suggest that their cognitive strategies for self-regulation are also less effective. We have studied this using basic executive function tasks, tests of cognitive control under conditions of stress, such as time or social pressure. For instance, in a verbal fluency task, subjects try to say in sixty seconds as many words as they can that either start with a certain letter or fit into a certain category. This task induces minor stress, having to perform for the experimenter and respond quickly without making mistakes. Critically, subjects are asked not to repeat any words. Accidental repetitions indicate a failure in self-regulation, like an intrusive thought that pops out before your mind can stop it. Our studies, and others like it, have shown that nightmare sufferers perform worse on tasks like this, which measure cognitive control.[20]

One explanation is that their minds' resources are caught up by negative arousal. We know that negative emotion actually restricts cognition, so it's as if the stress induced by the task short-circuits attention (it becomes harder to think freely when a threat crops up in our periphery, stealing our concentration). This seems to happen even within the nightmare itself. In the first place, the brain has less of a capacity for cognitive control in dreaming compared to waking life (our frontal cortex is relatively inhibited during REM sleep). Normally the neurochemistry of sleep helps to keep arousal in check, but in the case of nightmares, arousal is elevated, and the sleeping mind is ill-equipped to contend with the intensity of feelings. Within the nightmare, any available cognitive attention seems to be hijacked by a potential threat, and the threat becomes magnified as it draws more on the dreamer's focus and provokes an increasingly fearful response. The dreamer is pulled towards a fight, flight, or freeze response, beyond the realm of self-control.

To give one example from one of my nightmare subjects:

I was walking along the beach and it was raining lightly. The sun was setting over the ocean and I thought this was weird, as it was setting in the south and as I for some reason felt it was the middle of the day still. Suddenly it was very dark, there were no visible stars, I couldn't see the moon, and there weren't any streetlights. I didn't know which direction to walk in to avoid ending up in the sea. I started walking but slipped over and ended up in the water. I tried to swim to the surface but I couldn't figure out which way to go. My lungs were burning and I couldn't reach the surface, everything was still dark and I felt dizzy and felt my eyes close. I'm fairly certain I died.

Here, as in many nightmare examples, we can observe what's called an 'emotional cascade', a volatile downward spiral of narrowing attention and building negative emotion that limits our ability to respond usefully to a threat. This process can occur in waking life, but it is amplified in dreaming, where one's thoughts and emotions, and even the dream world itself, are engineered by the mind and manipulated by our feelings.

With nightmare sufferers, especially stimuli that are relevant to underlying adversity or trauma can cause alarm. A dream of seeing your partner talk to another character will be perceived as a threat if it's related to a history of maltreatment or neglect. While in waking life these external events are (relatively) unaffected by our inner dialogue, within a dream, our feelings actually change the dream's content, so the dream becomes congruent with our expectations. This creates a sort of feedback loop that quickly spirals out of control: dream content is perceived as threatening, the dreamer feels afraid of what might happen, the dream becomes even more threatening in response to the dreamer's feelings, and so on.

This is an extreme form of the pattern mentioned earlier, where feelings shape dream images. Here the intensity of feeling amplifies the vividness of the dream. At the same time, the dreamer's attention is restricted, narrowing in on and reinforcing a threatening image, to the exclusion of any other dream content. This pattern is evident in most nightmares, which typically end in the depiction of a singular overpowering and unrelenting threat – imminent death by an aggressor, a looming tidal wave. This is called a 'central image',[21] a singular image that evokes an overwhelming negative emotion; these images are common to nightmares and seem to grow more dominant and overpowering over the course of a nightmare, to the point of overwhelm and awakening.

So far we've seen how an intensity of negative emotion causes nightmares to develop. Going a step farther, nightmares seem to interfere with a natural dream process of adapting to stress. Recall that a function of REM sleep is to assimilate emotional experiences into memory – building new associations in memory and regulating the arousal attached to emotional memories. This process seems to fail in nightmares following adversity. Traumatic memories, for instance, often persist as vivid episodic memories that are easily recalled in their entirety and continue to cause significant distress even years after their occurrence. This may be in part a result of nightmares interrupting sleep's function.

One leading model of nightmares describes how during normal REM sleep, negative emotional memories are activated and recombined with other neutral memories, essentially remixing negative events with new and safe contexts.[22] This is a way to adapt to stressful experiences in the associative arena of dreaming; it's likened to a biological process of fear extinction, where fearful stimuli are paired with neutral or safe stimuli in order to overwrite a fear response. At a brain level, this process is supported by the amygdala, responsible for emotional arousal, and the hippocampus, where multitudes of contextual memories are stored. In this model, when negative memories are recombined with other neutral or positive memories, it limits the emotional response to the negative memories and gradually extinguishes or diminishes the fear associated with these memories.

We established this as a general purpose of REM sleep earlier, to break down memories into small fragments that are associated with similar memories from the past, gradually assimilating them into woven networks of related experiences. In the nightmare model, emotion regulation is key to this process: frontal brain regions are needed to limit amygdala arousal, so that the different negative memories can be accessed without causing overwhelm. This process fails in nightmares because of inadequate regulation of arousal, and the attempts to extinguish fear become unsuccessful. This is especially evident in the case

Why Dream at All?

of traumatic memories. It's as if the memory is too potent and intense emotions like panic, grief and rage that are associated with the event are unable to be quelled. The mind is overwhelmed by negative emotion and it becomes difficult to break down the memory and link it to others.

This interferes with the normal, helpful purposes that a typical dream might serve. Remember that in healthy dreams, there appears a free flow of connections, forward-backward-sideways, to loosely associated concepts, joining together unexpected elements of memory. It's usually as if REM sleep is temporarily lifting the 'Hebbian' highways of waking thought, allowing us to dissolve closely associated links (such as the features and context of a recent negative experience) and form new and unexpected, potentially adaptive links in their place (such as linking to an experience of safety or overcoming a similar situation in the past). This helps us to update memory over time and over experience. This is especially relevant when we need to *unlearn* a dominant association, like having a fear response to all buses after witnessing an accident. We need to be able to alter our response in the face of new and nuanced information, to realise that not every related scenario should cause the same fear to resurface.

With nightmares and trauma, we are unable to do this. A pattern of overfitting occurs – where any new related stressor triggers the same old trauma response.[23] This is especially evident in recurring nightmare scripts,[24] where we have the same stories repeatedly reinforcing a learned and unhealthy response to a particular threat (e.g. every time I see a bus in a dream, it evolves into a nightmarish accident). Nightmares, with their recurring emotions and content, seem to be stuck in a negative feedback loop. We become unable to break down a negative memory and allow for novel connections to form in its place; we are unable to *unlearn* a dominant stress-response pattern.

On the one hand, this is the very purpose of intense emotion in the first place: it flags an experience as vitally important for storage, asking us to hold on to a clear memory as a stark reminder to avoid similar situations in the future at all costs. But this quickly becomes maladaptive. The learned fear response becomes applied to any new and related stressors, and we become unable to let go of the trauma, instead drawing new experiences into its magnet over time.

Thus, in nightmares an overwhelming negative memory resurfaces again and again, without change or adaptation. In the absence of nightmares, dreams seem to provide a useful mechanism for adapting to stress: dreams seem to sift through memory banks, highlighting and recombining aspects of experience in helpful ways; and as memories become less emotional and less arousing, they can be assimilated into autobiography over time.[25] Nightmares extensively interrupt this sleep function; they showcase how this process of dreaming goes wrong.

This begs the question: Are nightmares always a normal function gone wrong? Or could they serve some health-related function, up to a certain point?

The question of whether, and to what extent, nightmares serve a function, that is, whether they offer some biological advantage, has been the focus of much theoretical research. The fact that nearly everyone has occasional

Why Dream at All?

nightmares suggests they are at least a normal response to stress. Large population-based studies show that 40 per cent of adults have at least one nightmare per month, and up to 85 per cent of people have at least one nightmare per year. Bad dreams, of course, are even more frequent, occurring up to four times more often than nightmares. Remember, bad dreams are less intense than nightmares and do not provoke awakenings. While thematically similar to nightmares, bad dreams are more likely to resolve positively, whereas nightmares end abruptly at peak negativity.

Most theorists agree that bad dreams are functional up to a certain point, which is a good thing, since the majority of dreams could be described as negative or 'bad'. By and large, theories propose that bad dreams serve an emotion regulation function, though the mechanism proposed, that is, *how* this is accomplished, is still up for debate.

One suggestion has been that dreaming helps with emotional problem-solving, where freely associating a negative event from waking life to novel ideas in the creative dream state gives rise to unexpected solutions to conflicts – solutions that are hard to come by in the more structured form of waking thought. This is similar to the view that pairing frightening experiences with neutral and safe contexts in a dream serves to extinguish fear. These two approaches take the same evidence – that dreaming highlights recent negative events along with distantly related memories – and suggest two parallel ways that this process might benefit emotional well-being: either through creative problem-solving or by alleviating fear through recontextualising.

If either or both of these theories are true, the next

question becomes: if the dream is intentionally highlighting our most upsetting waking experiences, how does the sleeping brain keep arousal in check, when confronted by these most unpleasant moments of our lives? In fact, this is one of the keys to emotion regulation theories, that although negative mental content is purposefully activated in the dream state, at the same time arousal is inhibited, controlled, subdued in some way. This is thought to be essential for the dream's function to occur. If negative emotions build up too much, the dream is fatally disrupted and we wake up. This is part of what goes wrong in nightmares.

From one perspective, the hyper-associative design of dreams helps to disperse negative arousal. An abundance of dream elements serves as a distraction, almost diluting a stressful memory by jostling it in with a handful of other curious and benign memories. In this framing, a dreamer does not become overwhelmed by negative arousal simply because they are distracted by the sheer wealth of alternate contents competing for attention. Most theories go farther, though, and describe the novel contents as meaningful rather than random, contents made of memories that are distantly and usefully related to one another. In this case, the associations not only serve as a distraction but also offer relevant contexts for reframing; an old memory suggests how the dreamer could better respond to a threatening scenario. If someone witnesses a bus accident, perhaps an old memory of a pleasant tour bus ride along the coast offers the dreamer an example of when buses were safe.

While these theories deal with the content of the dream, others describe how the body physically limits

Why Dream at All?

arousal during REM sleep. For instance, rapid eye movements themselves could have an effect similar to EMDR therapy (eye movement desensitisation and reprocessing therapy), where looking left and right quickly floods the brain with visual stimulation, which distracts from emotionally laden content. A similar somatic theory claims that the low levels of physical arousal during REM sleep relative to wakefulness means that in dreams we do not experience the same increases in heart rate or breath rate that we would expect during stressful events in wakefulness. This can serve as a form of desensitisation, where re-experiencing negative memories results in lower-than-expected levels of physical arousal, and the body in essence unlearns its stress response to the memory. The same mechanism could be said for the neurochemistry of REM sleep, where relatively low levels of noradrenaline (compared to wakefulness) mean there is less arousal in the brain despite the emotional content of dreams.

Whatever the mechanism, dampened arousal seems to equip dreamers with a greater capacity to confront (at times extraordinary) challenges and to explore how to respond to threats. This ties into an overarching evolutionary theory of dreaming, the threat simulation theory,[26] where the purpose of bad dreams is to provide a realistic virtual setting to practise dealing with the threats of waking life. Typical dreams of being attacked or chased (two of the most common bad dream themes) are considered prime examples of threat simulations that would provide a survival advantage to humans, preparing us to handle such scenarios in waking life. Over time this nightly rehearsal is thought to have given an evolutionary advantage to our species. While this theory focuses only on threat, similar arguments could be made for any of the scenarios we explore in dreaming, that the relative 'safe space' of the dream world allows us to test out challenging skills, difficult social relations, and so on.

Overall, theories abound on the precise way that dreaming accomplishes emotion regulation: through pairing negative content with novel or safe contexts, dampening bodily arousal through distraction or inhibition, and more. While these theories are often lumped together as 'emotion regulation' theories of dream function, they focus on different outcomes: from emotional problem-solving, to fear extinction, or rehearsal of threat coping skills. Generally, most theorists would agree that bad dreams help us to adapt to stressful experiences, up to a certain point.

Nevertheless, an excess of negative emotion (fear, sorrow, anger) and both brain and bodily arousal seem to interfere with this function over time. In chronic nightmares, repetitive and highly emotional content recurs that lacks the novelty and creativity of normal dreams. A dreamer reinforces again and again the same unsuccessful response to a threat and awakens in the midst of this failure. Without intervention, nightmares like this can become more and more frequent and disruptive to daily functioning. This cycle of pathological nightmares will be the focus of the next chapter. And later we'll see how the apparent balancing act of healthy dreams, actively confronting stressors while successfully managing emotion, is key to therapeutic interventions for nightmares.

*

Why Dream at All?

Overall, we can understand dreams and nightmares as very real lived experiences of the mind, brain and body. Dreaming at its minimum is a feeling, a subjective experience. Dreaming portrays a mix of emotions, sensory impressions and thoughts, and although dreams appear to be ephemeral and illusory, our feelings within and in response to dreams are impactful and have lasting effects on mood and memory.

Within dreams, feelings motivate us, and we are spurred to move our bodies and interact with the dream world. The available courses of action are less constrained than in waking life, and we can freely explore the felt consequences of our actions in the dream world. The dream world is also less constrained than the 'real' world, often morphing in response to our actions and emotions, and sometimes taking a turn for the worse, as in nightmares.

Over time, dreams and nightmares can change or reinforce how we remember and feel about the past. Although dreaming cannot change *what* has happened, it can change the *feeling* of what has happened. Dreaming does this anew each night, revising the shape of our autobiography, selecting what we choose to remember or to forget. In turn, this changes who we are in waking life, and how we perceive and respond to new experiences as well.

While nightmares are a natural response to stress, over time they can interfere with a function of sleep in adapting to adversity, disrupting how emotional memories are stored and holding on to associated distress. When unchecked, nightmares can become chronic, resulting not only in sleep loss but also waking anxiety and distress. In a way, a nightmare can act as a signal to the conscious mind

– it can tell us that something is wrong. This is perhaps one of the most functional uses of nightmares: offering a means of uncovering and working through emotional traumas within therapeutic contexts. In the next few chapters, we'll see more clearly how nightmares intersect with waking life and we'll explore the many methods available to treat nightmares, to restore sleep and mental health in turn.

4

When Are Nightmares a Problem?

Around the time I finished my PhD, I attended an international symposium on nightmares and nightmare treatment in Hanover, Germany, with a couple dozen other researchers from around the world. Over the course of the symposium, some of the key questions we aimed to answer were: who has nightmares, when, and what are the consequences?

Two of the attendees had recently published a paper that captured the importance of our cause: 'Nightmares: Under-Reported, Undetected, and Therefore Untreated.'[1] Nightmares were slipping under the radar of modern medicine and public awareness, despite by now robust evidence that they critically impact health.

Already in 2016, the research was conclusive that nightmares are a serious problem for some people, with severe consequences on mental health including increased depression, anxiety, insomnia and even suicide risk. But there was very little awareness in the broader health care field on how to diagnose and deal with nightmares – a problem that persists to this day. Starting that summer in Hanover, and in many collaborations since, we have worked to develop a consensus on the best clinical approaches to assessing and treating nightmares.[2]

First and foremost, it became important to establish clear guidelines on how to identify when, and in whom nightmares become a pathological problem, diagnosed as nightmare disorder. For instance, we know of several personality traits that make someone more vulnerable to nightmares and to the harmful effects of nightmares. Second, although nightmares are clearly emotionally distressing, we wanted to get the message out that nightmares can also damage other areas of mental and physical health, not to mention sleep health. Moreover, having nightmares increases your risk of developing other health disorders later, especially PTSD, and even cardiovascular disease.

Nevertheless, one of my main contributions to the field has been to highlight the potential upside of being nightmare-prone; that those of us who are susceptible to nightmares are also more sensitive to positive experiences in waking life and have a more vivid and impactful dreaming life altogether. As a complement to my work on nightmare disorder, this research is starting to clarify some of the possible benefits of being prone to nightmares and also points towards new avenues of intervention.

Who develops nightmare disorder

For whom do nightmares become a problem?

According to official guidelines, 'nightmare disorder' is a clinical condition, where having at least one nightmare per week severely interferes with quality of life.[3] These nightmares typically occur in the latter half of the night, and sleep studies suggest they especially occur during

When Are Nightmares a Problem?

morning REM sleep (though they can occur at other times) and are clearly recalled upon awakening. A total of 4 to 6 per cent of the population meets these criteria, with weekly nightmares that cause significant waking distress.

The intense negative emotions felt in nightmares most commonly include fear but can also include anger, sadness, or grief. While in their milder form, bad dreams seem designed to help us respond and adapt to stress, nightmares become pathological when they occur too frequently and cause too much distress. In fact, frequent nightmares often co-occur with other waking mood symptoms, such as depression and anxiety, which suggests a breakdown in healthy emotion processing.

One of the key factors in who develops nightmare disorder is how much distress their nightmares cause – how the overwhelm in the dream spills over into waking life.

The amount of distress someone experiences in response to nightmares varies widely from person to person and in part determines whether nightmares develop into a clinical condition. Where one person might simply laugh off a nightmare or feel a sense of relief or curiosity upon waking, another might worry about a nightmare all day long, unable to shake it. When we measure nightmare distress, we look for several different indicators, since it can manifest in many ways. The most obvious is an immediate emotional reaction to the nightmare, but patients might also struggle with repetitive thoughts about nightmares during the day, or have difficulties coping with nightmares, and feel like nightmares harm their well-being. Nightmares can interfere with other areas of life such as relationships or work, making it difficult to focus and get through the day. One of my subjects said that during periods when she was having nightmares almost every night, she dropped out of school and found it difficult to even leave her bedroom during the day.

These various forms of nightmare distress are often dealt with through unhealthy or unhelpful coping mechanisms. The most common is an attempt to avoid nightmares: to try to suppress the recall of a nightmare, and avoid going back to sleep to evade further bad dreams. Nightmare sufferers also feel helpless, that nightmares are beyond their control. All of these forms of distress lead to a downward spiral, where nightmares worsen daytime stress, resulting in further and worsening nightmares over time.

It seems like the distress caused by nightmares is more important than the sheer number of nightmares people experience, and this distress is more closely tied to mental health consequences of nightmares, which we'll see later. That said, of course the more frequent nightmares are, the more distressing they can become.

While most individuals do not have nightmares regularly (and occasional nightmares tend not to be a cause of concern), several personality traits can predispose someone to having more frequent and distressing nightmares. In the first place, age and gender play a role, with women and those of a younger age more prone to nightmares. In general, both women and younger adults recall more dreams, which may explain their propensity for nightmares as well. Nightmares (and dreams generally) become less frequent in older adults starting around forty

When Are Nightmares a Problem?

years of age, which is possibly due to changes in sleep structure (less REM sleep) that occur, or other changes in cognition and emotion over one's lifetime.

The personality trait of 'neuroticism' has been widely studied in relation to nightmares; this somewhat antiquated term describes a tendency towards negative mood states such as anxiety or irritability alongside frequent feelings of guilt, envy and shame. Individuals high in neuroticism are more sensitive to stress and display a sort of emotional instability; everyday situations are seen as more serious and threatening and can cause volatile mood swings. In brief, neuroticism exaggerates emotional reactivity, which leads to more nightmares (in response to heightened waking stress) and to more distress felt in response to nightmares.

In addition, the basic physiological trait of 'hyperarousal' refers to having habitually high levels of bodily arousal, even during rest, and is common in nightmare sufferers. This heightened bodily arousal means that emotional events, including nightmares, take a higher physical toll and can easily become overstimulating and overwhelming.

A broader personality trait aptly called 'nightmare proneness'[4] groups together elements of neuroticism and hyperarousal, along with other particularities common to nightmare sufferers including some indications of paranoia, and a tendency to somatise stress symptoms, that is, to experience bodily symptoms such as chest pain or headaches in response to psychological stress. This unique combination accurately predicts one's propensity to distressing nightmares.

Into the Dream Lab

At the same time, some personality traits associated with nightmares seem to be more positive, like openness to experience, which describes individuals who are highly receptive to emotional and perceptual experiences and report more pleasurable responses to positive and aesthetic experiences. This trait overlaps substantially with a physiological trait termed 'sensory processing sensitivity', which describes individuals who are more sensitive to sensory and emotional stimuli and are highly attuned to their environment. These individuals can be generally defined as being 'highly sensitive', with a vivid experience of the external environment and a rich and intense inner life, too. (We'll explore this more later in the chapter.)

Together, these factors of emotional reactivity, bodily arousal and perceptual sensitivity all correspond with a propensity to nightmares. In addition, there are several cognitive traits that predispose individuals to nightmares.

For instance, avoiding certain thoughts during the day is paradoxically related to their intrusions in nighttime dream content. As we saw earlier, people who attempt to suppress unwanted or undesirable thoughts during the day also dream more about these problematic thoughts come night. More generally, so-called thought-suppressors – people who habitually suppress unwanted thoughts – also have dreams that are more related to daytime emotions and relationship worries. Presumably, this is a result of a dream 'rebound', that is, the return of these evaded worries into the mind once a person is asleep and dreaming.

In a similar vein, psychologist Victor Spoormaker claims that avoiding thoughts of nightmares themselves can perpetuate their recurrence.[5] It is very common for

When Are Nightmares a Problem?

nightmare sufferers to try to avoid thinking about nightmares they have just awoken from, such as immediately getting out of bed or trying to distract themselves or clear their mind of a nightmare. For Spoormaker, this cognitive avoidance interferes with our ability to deal with or approach head on the issues causing the nightmare, and thus spurs its recurrence. In the absence of any efforts to manage or resolve the nightmare, it becomes a sort of habitual 'script' that replays whenever triggered by stress.

While avoidance is a common crutch used on awakening, many nightmare sufferers are wrought with worry at the end of the day, lying awake in bed and filled with dread before falling asleep. This is called 'presleep cognitive arousal'; it is common in nightmare sufferers and also in insomnia, and it prolongs the amount of time it takes someone to fall asleep. Recently it was shown that the amount of presleep cognitive arousal on any given night actually predicts on which night a nightmare will occur.[6] This finding is revelatory for trauma survivors, who experience frequent but sporadic and unpredictable nightmares – and not knowing when a nightmare will occur is itself a huge source of anxiety that could (ironically) incite more nightmares.

In trauma survivors, as we'll see later, the content of thoughts swirling around in the mind prior to sleep can also act as 'reminiscent stimuli' – that is, stimuli that remind an individual of a traumatic event. While any type of presleep cognitive arousal can disrupt sleep, such as worrying about work or financial problems, reminiscent stimuli are more directly related to a traumatic event, such as remembering the people or places involved in the trauma. Even in waking life, reminiscent stimuli can induce flashbacks, which are akin to waking nightmares, where unwanted memories of a trauma intrude into the mind and make a person feel as if they are reliving the memory. Reminiscent stimuli can similarly trigger nightmares when encountered just prior to sleep.

Besides presleep cognitions, more general cognitive traits such as a tendency to dissociate or to distance oneself from reality to avoid pain, and having more disorganised thoughts, have been correlated with nightmares. And those who have difficulty identifying or describing their feelings – termed 'alexithymia' – have more nightmares. These patterns of distancing from or struggling to confront one's emotions are aggravated by hyperarousal; in other words, when emotions are running high, it feels impossible to discern or to sort through them. As you can probably guess, these traits are unhelpful for managing emotions, an overarching problem faced by nightmare sufferers.

Even waking daydreaming style can predict who has nightmares. Research has identified three broad styles of daydreaming, and people tend to fall into one style more than others. Positive constructive daydreaming is a form of active, enjoyable and creative mental imagery, often of a wish-fulfilling fantasy genre; individuals with this style of daydreaming do not have many nightmares. Those with guilty-dysphoric daydreams, on the other hand, have mental imagery that is often negative and failure-oriented, featuring guilt and hostility. One example is the 'suffering martyr' daydream, an imagined scenario where friends or family members express regret at not acknowledging your positive or impressive attributes in the past. Those with more dysphoric daydreams, unsurprisingly, have more nightmares. The third daydreaming style, poor attentional control, describes people who have unintentional and uncontrolled mind-wandering that can be absorbing, in contrast to people with high attentional control who are able to focus and direct their inner life. Of course, it's those with poor attentional control who have more nightmares. In some cases, daydreams can become so out of control and dysphoric that they are more aptly termed 'daymares', taking on an experiential quality similar to nightmares.

All that said, some of these cognitive patterns, such as dissociation or absorption (becoming absorbed in fantasies), do have an upside, being related to increased creativity and imaginativeness. This paradoxical relationship is a recurring one in nightmare research, where the same traits that predispose individuals to nightmares can also have advantages, including increased creativity, and also empathy and aesthetic sensitivity, which we'll explore in more detail later.

Stress, trauma and the case of PTSD

We know who is vulnerable to nightmares, but when do they become a problem? What causes a more pathological pattern to emerge?

Of course, nightmares are most likely to become problematic when someone is exposed to high levels of stress – such as changing jobs or going through a divorce – or in more traumatic situations like living in a war zone. As one example, nightmares increased in the United States after

9/11, and the content of these nightmares sometimes clearly related to the attacks, such as planes being hijacked, crashing into buildings and bombs. Other studies of prisoners of wars and Holocaust survivors have found war-related nightmares to be common in these individuals, and these themes can persist for years. For example, even decades after World War II, veterans continued to have frequent nightmares related to their war experiences. Natural disasters also cause nightmares, such as hurricanes or wildfire; in one study, survivors immediately affected by such disasters had nightmares as a result, but not others who lived nearby and were not directly affected.

The COVID pandemic also marked a clear case where collective stress impacted dreams worldwide. Universal concerns, such as the threat of contagious disease, social isolation and lockdowns, and personal and economic uncertainty, all emerged during the pandemic. Numerous studies reported changes in dream content during this time, including a surge in nightmares during lockdown and nightmares being associated with COVID-related stress, such as having concerns about contracting COVID, having infected family or friends, or reading news about the virus. Studies across Brazil, the United States, Italy, Canada and China observed general increases in bad dreams, with themes of anxiety, sadness, aggression and preoccupations with health and death. Some of these themes were clearly related to the pandemic, such as failures in social distancing or vaccine fear. Other themes that existed before the pandemic but became more pronounced in this time period include dystopian and apocalyptic dreams or dreams of failure and death.

When Are Nightmares a Problem?

In a Canadian sample,[7] the most common dream theme during the pandemic was of inefficacy – trying again and again to do something but failing. This surpassed the number one pre-pandemic theme of 'being chased or pursued'. Although these are both typical dream themes that existed pre-pandemic, other novel themes emerged that more directly reflected pandemic concerns, such as 'being separated from a loved one'. This theme reflects the challenge of social isolation, one of the most pressing concerns at the time. Other dreams were more health-related (dreams of hospitals, being sick, germs and contamination), or featured generally intense emotions like fear, anger, hopelessness and loneliness. These new dream patterns emerged as, quite unprecedentedly, concerns about COVID spread around the world.

While the pandemic revealed the impact of global stress on dreams, it also resulted in a worldwide increase in post-traumatic stress disorder (PTSD), which develops in some individuals after experiencing traumatic events. PTSD diagnosis is based on symptoms of re-experiencing a trauma in intrusive flashbacks or nightmares, having generally heightened arousal, and using unhealthy coping mechanisms like emotional numbing and avoidant behaviours to try to lessen symptoms. Remember that re-experiencing occurs when the memory of a trauma is involuntarily recalled, usually due to cues in the environment that are associated with the trauma (i.e. reminiscent stimuli). The resulting intrusive memories – or flashbacks – can be remarkably vivid, intensely emotional, and the patients really feel as if the trauma is happening right then and there.

Post-traumatic nightmares are a form of re-experiencing where traumatic memories are replayed during sleep, and these are considered a 'hallmark' of the disorder.[8]

Up to 90 per cent of PTSD patients experience nightmares, as often as six nights a week. In more than half of these cases, nightmares take on a 'replicative' form, meaning they clearly incorporate elements of a trauma or even replay it almost exactly. Replicative nightmares are the most severe and pathological form of nightmare and can persist for the rest of one's life, perpetually triggered by reminiscent stimuli encountered in waking life or even within dreams. Some authors have claimed that these nightmares 'themselves could be traumatic for the dreamer'.[9] More generally, PTSD patients have high levels of aggression and anxiety in their dreams and frequent scenes of death and violence. Of course, post-traumatic nightmares cause significant distress for patients, and they are associated with more severe waking PTSD symptoms, too.

To demonstrate just how severe and lasting the consequences of trauma can be: Holocaust survivors, who were victims of extreme and prolonged physical, mental and emotional violence and stress, still experienced considerable PTSD symptoms and nightmares even forty-five years after liberation.[10] In Holocaust survivors, the frequency of nightmares even this many years later was directly proportional to the duration of internment in concentration camps; those exposed for longer periods continued to suffer more for decades, trapped in the mental prison of nightmares.

In the months following liberation, one Auschwitz survivor reported:

When Are Nightmares a Problem?

In my dreams, I saw bloodied washbasins (they always terrified me) on the block's ground floors, exercise routines in the yard, and prisoners killed with clubs, blood running down the gutter after the execution. I also often dreamed about an elderly couple standing in the yard in front of block 11, holding a sign reading 'because our son escaped,' etc. I always woke up drenched in sweat with my heart racing – scared and unaware of what was real.

Even decades later, another writes:

Our nights are no longer ours. Our dreams are still there, and those horrible episodes we experienced continue to haunt us, robbing us of well-deserved rest. We are the ones who see that reality nearly every night, right before our eyes. It transforms our sleep, forcing us to continuously revisit the alternative, brutal endings to our lives – the deaths we once avoided. The death sentence we received from people now long gone has been postponed, but nothing, nothing can make their evil small enough to free us from the nightmares.

Polish researcher Wojciech Owczarski describes these persisting nightmares as unrelenting testimonies to the brutal experience faced in concentration camps decades earlier. One survivor described being tormented by a repeating dream of *'being shot by a firing squad; I see the entire ceremony, they tie my hands. Unable to defend myself or avoid danger in time, I wake up as the bullet penetrates my skull.'*
Long after liberation, these nightmares persisted.

Perhaps counterintuitively, a parallel long-term consequence of PTSD is a gradual loss of other types of dream recall.

This has been described as an innate mechanism for coping with severe trauma, a form of dream (and nightmare) repression. For instance, in Holocaust survivors, those who were better-adjusted in the long term remembered fewer dreams when awakened from REM sleep in the lab (33.7 per cent) than less-adjusted survivors (50.5 per cent) and control subjects (80 per cent). Well-adjusted survivors also had less complex and less salient dreams and often reported having no impression of dreaming at all.

This diminished dreaming could be a kind of adaptation, designed to suppress the occurrence of nightmares. Some survivors said outright that they wanted to not only forget about their nightmares but deny their dreams altogether:

> I want to forget my dreams as quickly as possible. I see nothing special about them. I am aware of the factors that cause them. I do not read books about dreams, and I never try to explain them. Sometimes, after a day filled with films or books, I need to prepare myself mentally for the cursed midnight cinema.

This aversion to dreaming, and to sleep more generally, is often reported by nightmare sufferers and especially those with PTSD and is sometimes part of a broader phenomenon called 'fear of sleep'. Patients are afraid of having nightmares, afraid of re-experiencing traumatic memories,

When Are Nightmares a Problem?

and try to avoid nightmares and sleep altogether. In addition, some patients feel afraid of the loss of vigilance that accompanies sleep, meaning they feel anxious about not remaining watchful during the night or in the dark. This seems particularly true for veterans, who may have been trained to stay alert and on guard during the night.

Along similar lines, in other traumatised populations, having to stay awake or on call at night exacerbates nightmares and sleep problems. For example, firefighters experience a lot of physical and mental stress in their work, and they are repeatedly exposed to trauma, such as witnessing accidents and having frequent close brushes with death. Nearly 20 per cent of firefighters experience frequent nightmares with a high risk for PTSD, and this is aggravated by the necessity for shift work, with rotating schedules and irregular sleep patterns. Combining trauma exposure with irregular sleep compounds the risk for nightmares and PTSD.

Finally, while PTSD research has mainly focused on victims and witnesses of trauma, another group of people who experience frequent nightmares are those with 'perpetrator trauma' – individuals who develop a similar psychological profile to PTSD as a result of inflicting violence or pain on others. Studies among gang members, or even in the military or among police officers, have revealed that causing serious injury or death to another person leads to several symptoms comparable to PTSD. Perpetrators of violence report a constant re-experiencing of their offence through nightmares and haunting repetitions of their actions in intrusive imagery. Individuals who have committed murder report dreams marked by intense guilt; some even dream as if they are the victim of their own crime, reliving the anguish and helpless state of their victims. These nightmares occur in tandem with a set of symptoms including poor sleep, avoidance behaviours and generally poor mental health – all similar to PTSD.

All told, nightmares clearly occur as a result of waking stress and trauma, and in more severe cases this leads to replicative nightmares and PTSD. But chronic nightmares may be only the tip of the iceberg in terms of poor sleep and mental health. Even on nights without nightmares, patients with nightmare disorder experience fear of sleep and have difficulty falling asleep, along with shorter and worse sleep quality, and all of these things can have immediate and long-term consequences on health, as we'll see next.

Health consequences of nightmares

Unsurprisingly, chronic nightmares have numerous negative consequences on waking mental health and mood.

Perhaps most important, nightmares are a highly significant risk factor for suicide; they are associated with more thoughts of suicide, more suicidal behaviours and higher rates of death by suicide. Expert nightmare researcher Michael Nadorff has shown time and again that nightmares, more so than other factors such as anxiety or depression or insomnia, strongly predict suicide risk. The severity and duration of nightmares is directly relevant to this link: the longer a person suffers nightmares, the higher the suicide risk. Moreover, nightmares lead to a fourfold increase in repeat suicide attempts and a twofold increase in actual death by suicide.[11]

When Are Nightmares a Problem?

In patients with PTSD, replicative nightmares pose the greatest risk for suicide. One of our large-scale studies of veterans showed that those with nightmares that incorporated memories of a trauma were more likely to reattempt suicide than those with unrelated nightmares. Another study of 16,220 hospital workers from Wuhan, China, found that 13 per cent of surveyed individuals reported suicidal ideation early in the COVID pandemic, and in those with trauma exposure, nightmares fully explained the increased suicide risk – not insomnia, depression, or anxiety.[12] Suffice it to say, assessing for and treating nightmares should be an essential component of suicide prevention programs, but the medical system is still lagging behind, as most health care providers do not ask patients about nightmares, nor do they treat them, and most nightmare sufferers are unlikely to report them spontaneously.

Nightmares have also long been known to correlate with other mental health problems. They are especially linked to increased depression and anxiety, and the content of nightmares is relevant to these links. For instance, during the pandemic, people who had more bad dreams of inefficacy also had more depression symptoms. Dreams of inefficacy seem to parallel some of the waking symptoms of depression, such as feeling helpless and incapable of reaching one's goals. At the same time, dreams of seeing people who were ill or dying (common themes during the pandemic) were associated with more anxiety and depression. The higher prevalence of this theme during the pandemic may have reflected real anxieties from waking life, faced with the uncertainty of physical health and possible death during this time.

In addition to immediate impacts on mental health, nightmares also increase the chances that someone will go on to develop a psychiatric disorder in the future.

Most notably, having frequent nightmares increases the risk that someone will later develop PTSD. As we saw earlier, even mild nightmares seem to develop in response to adversity and often feature recurrent, dysphoric themes related to adverse memories. In PTSD, this pattern is exaggerated: severe and replicative nightmares develop in response to traumatic events, a more extreme form of adversity. So there is a spectrum of nightmare pathology, where the severity and 'replicativeness' of nightmares reflects a relative response to the severity of waking life stress. When someone who already has recurring nightmares is exposed to a significant trauma, there is an increased likelihood they will develop PTSD, compared to someone who has never struggled with nightmares. Basically, the underlying pathology is already there, and the threshold to tipping the scales into full-blown PTSD is lower. Because of this, treatment of milder cases of nightmare disorder could actually protect against or prevent later PTSD.

Moreover, the content of nightmares in the immediate aftermath of trauma is predictive of who will go on to develop PTSD. We looked at dream content in the one-month period following hospitalisation for a trauma and discovered that those individuals who had more aggressive and negative dreams were more likely to be diagnosed with PTSD a month later. This is an important direction for further research, because it indicates there is a window of opportunity following trauma when proactively treating nightmares could prevent PTSD. In a similar study by my

When Are Nightmares a Problem?

colleagues in Rochester, about half of traumatised patients reported dreams that were 'highly similar' to their traumatic event in the month following the event, and these same patients went on to develop PTSD. The other half of patients either had dreams that did not depict trauma memories or they did not recall dreaming. These studies highlight that dream content that is directly or thematically related to a trauma, early after exposure, predicts who will develop PTSD, and these individuals would likely benefit most from early intervention.

To tie this back to the emotion regulation function of REM sleep and dreaming, it may be that increased negativity and arousal in the critical period following trauma disrupts REM sleep function and interferes with adaptation to this stressful experience. Indeed, other studies have found that measures of bodily arousal in the weeks following trauma, such as higher resting heart rate, predict subsequent PTSD. Thus, treating nightmares and managing arousal in the early aftermath of trauma could prevent PTSD. These treatments would be especially important for patients with a history of nightmares, who are presumably at the greatest risk.

Besides immediate links to mental health, chronic nightmares may also adversely affect physical health, notably cardiovascular health, in the long term. For instance, one large-scale study of US military veterans found that frequent and severe nightmares were associated with more self-reported cardiovascular disease, and this was independent of other risk factors like sleep quality, smoking and depression diagnosis.

The impact of nightmares on cardiovascular health could be mediated by the bodily arousal associated with many nightmares: when someone has a nightmare, their heart rate and breathing rate often quicken in the last minutes before awakening. In general, people who have chronic problems with physical arousal, such as elevated resting heart rate, are also more likely to develop cardiovascular problems. Both PTSD patients and nightmare sufferers without PTSD exhibit this type of hyperarousal during wakefulness and in sleep, and this could increase their heart health risks over time. (Other sleep disorders, such as sleep apnoea, are strongly associated with cardiovascular disease, and nightmares, too.)

Aside from the short-term effects of a nightmare on arousal, there are some more lasting effects of nightmare stress on the body. For instance, the stress hormone cortisol usually peaks about forty minutes after awakening and then decreases across the day. In individuals with nightmares, cortisol is more elevated in the mornings following nights with nightmares, although these effects disappear throughout the day. Nightmare sufferers also report more physical health complaints and lower energy resources on days following a nightmare. Altogether, in addition to the harmful effects of nightmares on mental health and mood, it seems likely that nightmares damage physical health and the body's stress response system as well. Clearly, nightmares themselves disrupt sleep with bodily arousal and induce a prolonged stress response on awakening. Both of these effects could harm cardiovascular health in the long term.

In sum, chronic nightmares are associated with deteriorating health, especially anxiety, depression, suicide risk,

When Are Nightmares a Problem?

PTSD, poor sleep, fear of sleep and increased physiological stress and cardiovascular risk. Altogether, this research highlights the importance of treating nightmares in order to reduce physical health risks and resolve sleep and mental health problems as well.

The surprising upside of nightmares

At this point it is probably clear that chronic nightmares are harmful to multiple facets of health. And indeed, most research on nightmares has focused on the negative causes and consequences of their occurrence. A common way of viewing nightmare disorder is within what's called a 'diathesis-stress' framework – where a diathesis is a trait that makes someone vulnerable to stress and, in this case, more likely to develop nightmares in response to stress. More broadly, this is a 'trait-state' model that looks at how certain traits (personality, physiology) interact with certain states (stress, trauma) to determine one's health.

The picture presented so far seems to coincide with a diathesis-stress model of nightmares, for example, that individuals high in neuroticism or hyperarousal have worse nightmares and mental health outcomes, especially when combined with a history of adversity.

However, diathesis-stress research is often biased, looking only at what happens to someone under conditions of stress. An alternative framework, called 'differential susceptibility', looks at how someone responds to both negative and positive conditions and describes traits that are 'for better and for worse',[13] that is, associated with either negative or positive outcomes depending on the situation.

Recent studies inspired by this framework have shown that many 'diatheses' – traits assumed to be associated with vulnerability to stress – actually afford some benefits under the right conditions. For instance, young adults high in stress reactivity (a genetic trait associated with heightened responses to stress) report greater stress on days when they have to take an exam, but lower stress on non-exam days, than those who are less reactive. In other words, they are better off than others when in a benign or non-stressful context.

My colleagues and I think that nightmare sufferers are best characterised by a trait termed 'sensory processing sensitivity';[14] this is a trait where emotional arousal (both negative and positive) amplifies one's sensitivity to the environment altogether. This trait develops in certain people in response to early adversity. If we remember, early experiences of adversity lead to increased reactivity in the limbic system of the brain and to less regulation of the amygdala. Most research has focused on how this leads to greater negative emotional reactions, but we now know that the amygdala responds just as strongly to positive stimuli as it does to negative. In other words, increased positive emotions, too, can result from these brain changes.

This opens up possible benefits to being nightmare-prone, and indeed there is building evidence that nightmare sufferers are more reactive to positive emotional experiences, in addition to their sensitivity to stress. In basic experimental studies, we have found that they rate both negative and positive images as more arousing than their non-nightmare counterparts, and they even outperform others in certain cognitive, perceptual and

When Are Nightmares a Problem?

social skills. In theory, having stronger emotional reactions leads to increased processing of the environment: the world becomes more salient, and so 'highly sensitive' individuals are more aware of and responsive to their surroundings, good or bad. This can be adaptive at times, for instance, becoming subtly attuned to emotional, perceptual, or social situations. But this also leads to being more easily overwhelmed, especially when in a highly stimulating environment.

This is similar to a personality trait that has been used to describe nightmare sufferers in the past, termed 'boundary thinness'.[15] From hundreds of patient interviews, clinician-scientist Ernest Hartmann was one of the first to draw connections between nightmare sufferers and sensitivity. He described nightmare sufferers as having 'thin boundaries': low physical thresholds for sensory stimuli, high reactivity to emotional stimuli, and increased fluidity between thoughts and feelings. He describes a general emotional and perceptual sensitivity and an increased attention to one's own and others' thoughts and feelings.

We and others have consistently found nightmare sufferers to have thinner boundaries and to be more highly sensitive than those who do not have nightmares, as scored on simple self-report questionnaires.[16] Subjects report that they are often aware of subtleties in the environment; easily overwhelmed by strong sensory input; affected by others' moods; sensitive to caffeine, bright lights, or loud noises; and appreciative of aesthetic sensory experiences, such as scents, tastes, arts and music.

But could nightmare sufferers really have a more sensitive body, one that registers stimuli more acutely? Do they really feel the world more deeply, or do they only *think* that they feel it more deeply?

Highly sensitive persons, at least, do have a more sensitive nervous system. They are better able to detect subtle changes in the environment than less sensitive individuals. For example, they perform better than others when tasked with detecting slight differences in landscape photographs. And they have more activation in brain areas linked to attention and perception during this task, too, suggesting an increased capacity for visual attention.

Early in my career while working in the UK, we tested whether highly sensitive individuals are also better at detecting subtle auditory stimuli. We used the phantom word illusion, where two-syllable words (e.g. No-Where) are repeated at four times the normal speed over two stereo speakers, alternately, for two minutes.[17] Presented in this way, the speech becomes ambiguous, and subjects sometimes have the illusion of hearing words that are not really there. This is called 'pareidolia', perceiving objects where there are none (like seeing shapes in the clouds). In the phantom word task, we asked subjects to report out loud any words that they could discern while listening. Highly sensitive subjects were better at detecting real words than less sensitive subjects, and they did not make up more illusory words. Said otherwise, they in fact do have more fine-tuned auditory perception.

Highly sensitive persons are also more responsive to social aspects of the environment; they have increased activation in several brain regions (insula, inferior frontal gyrus and cingulate cortex) when viewing others' facial expressions. These brain regions are involved in mirroring and empathy, suggesting that highly sensitive individuals are more attuned to the emotions and expressions of others. This appears to be the case for nightmare sufferers, too, who report more frequent mirror behaviours, such as contagious crying or laughing, and score as having higher empathy on self-report questionnaires.

Having greater social attention and empathy has clear benefits for promoting social connection and enhancing social reward (in other words, because one is more attuned to social situations, they will also reap more benefits from positive interactions). Clinical studies have shown that highly sensitive individuals respond well to social interventions, such as group treatments for depression that promote social connection. Similar group interventions could be effective for nightmare sufferers, who may respond especially well to positive social support during treatment, which we'll explore later.

Overall, those who are prone to nightmares appear to be highly sensitive to emotional, perceptual and social cues in their environment. Whereas 'diathesis-stress' models describe only the negative causes and consequences of nightmares, 'differential susceptibility' allows us to better understand this possible upside to being nightmare-prone.

Given what we know about how feelings give rise to dream content, it might not be surprising to learn that nightmare sufferers have generally intensified dreams and a vivid inner life, including frequent positive dreams (although this is still a surprise for many clinicians who are stuck in a diathesis-stress view of nightmares). Research has shown that those with nightmares have more vivid, bizarre and even intensely positive dreams and daydreams. They remember more dreams in general, report more lucid dreams, and have longer and more immersive dreams than their non-nightmare counterparts. In our studies, we have found that nightmare subjects recall on average three to five non-nightmare dreams each week (that's in addition to their weekly nightmares). This means that most of their dreams are *not* nightmares, and even these dreams are more frequent, vivid and emotional than is usual.

Highly sensitive persons, too, are known to have rich 'inner worlds' – a complement to being sensitive to the external world; they report especially vivid dreams, more disturbing nightmares and more lucid dreams, too. Earlier work on 'thin-boundaried' individuals also found them to have more vivid, frequent and bizarre dreams and nightmares, along with immersive daydreams and absorbing fantasies. In lucid and archetypal (vivid and mythic) dreams – more common in thin-boundaried individuals – the imagery is often euphoric and displays a sensory vibrance and narrative elaboration uncommon to typical dreams. We can think of these dreams as the lunar twins of nightmares: with intensely positive, perceptually rich and cognitively powerful dream content.

In a way, the dreams of nightmare sufferers parallel waking features of sensitivity; they are more strongly emotional, perceptually vivid and socially engaging.

Even the bodies of nightmare sufferers are more sensitive while asleep. In the first place, we have found that their dreams and daydreams contain more body sensations, both pleasant and unpleasant. They are also more sensitive to real sensory stimuli during sleep, more often

When Are Nightmares a Problem?

perceiving external cues such as flashing lights or beeping sounds within dreams. There is even evidence that they are more sensitive to internal body signals during sleep. One study looked at the 'heartbeat-evoked potential', which is a measure of the brain's processing of the heartbeat. This is a kind of interoception, a sensing of one's internal body states. Nightmare sufferers have a magnified heartbeat-evoked potential during REM sleep compared to healthy subjects, meaning their brain is more responsive to this subtle bodily signal while asleep. A sensitivity to real body sensations during sleep could in part explain the vivid sensory (and emotional) dreams they so often experience.

So how does such vivid dreaming impact dream function? Does it have any benefit, or is it disruptive, like nightmares?

We know that in waking life, positive emotion expands cognition; it broadens mindsets, increases openness to others and encourages exploration, curiosity and creativity. Similarly, in sleep, intense positive emotions can conjure vivid, elaborate and meaningful dreams. I mentioned that nightmare sufferers have more lucid and archetypal dreams – these are often experienced as euphoric, insightful and altered state experiences. These rich and aesthetic dreams can then positively impact waking life; they can fill dreamers with awe or pleasure, spark creativity, or even elicit spiritual feelings or self-reflection, both within dreaming and on waking.

That these positive states can carry over into waking life is in direct contrast to the distress caused by nightmares. We know that on mornings after nightmares, subjects report more anxiety, sadness and even physical pain than on other mornings. But positive dreams promote feelings of exhilaration, vigour and generally positive moods in the morning. Positive dreams also correlate with waking peace of mind, and even with better subjective sleep quality. When we tracked subjects over eight weeks of keeping daily dream and mood diaries, we found that on mornings after pleasant dreams, subjects reported a better mood and, besides emotion, other features of dream content were relevant, too. By analysing the text of dream reports, we found that nightmarish dream elements, such as references to death, were linked to more negative mood in the morning, whereas dreams with more social words were linked to better morning mood. We'll see later how curating positive social content, and other helpful themes in dreams, could actually support patients in overcoming nightmares.

Finally, cognitive features of dreaming also impact how we feel on waking. Dreams high in mindfulness and self-reflection – where we pay attention to what we are sensing and feeling within the dream and even reflect on our experience as it occurs – promote self-reflection following sleep. When dreams are more realistic, they promote stronger self-perception, a better image of oneself, on waking. Archetypal dreams, too, are often deeply meaningful and invite self-reflection and spiritual contemplation in the days after their occurrence.

All of these examples present a positive counterpart to post-awakening nightmare distress and offer a direct benefit of pleasant dreams on sleep and mental health. Once we start to delve into the specifics, it starts

When Are Nightmares a Problem?

to become clear that this link is due not only to dream emotion but also to social, bodily and cognitive features of dreams. These findings begin to reveal when and how dreams can be beneficial or harmful to waking mental health, and – crucially – where we can intervene in the case of nightmares.

One of the more important examples of this comes in the form of lucid dreams, where you become aware that you are dreaming while still asleep. Lucid dreams are linked to immediate benefits, such as feeling more refreshed and joyful in the morning. Of course, lucid dreams themselves are often positive because dreamers choose to engage in fun activities like flying or breathing underwater – exciting acts impossible in the real world. Subjects report using lucid dreams to overcome nightmares, fulfill wishes from waking life, and even promote emotional and physical healing. It's worth noting that some of the benefits of lucidity seem to be independent of how positive the dream is. In other words, regardless of whether a lucid dream is enjoyable (a usual goal of lucid dreamers), the simple fact of having some agency – a little bit of insight or control in dreams – seems to benefit waking mood. In this way, teaching people how to have more agency and lucidity in dreaming could support mental health, and we'll dive into this more when exploring lucid dreaming as a treatment for nightmares.

In general, there is growing evidence that dreaming strongly predicts how we feel in the morning, often even more so than other measures of sleep (such as sleep duration or sleep quality). Nevertheless, there is little research on the potential long-term benefits of pleasant dreams. It is possible that positive dreams can improve well-being over time in a manner referred to as 'upward spiral dynamics',[18] where positive dreams prompt us to go into the day with more curiosity or engagement, which then leads to further positive waking experiences, and better sleep and dreams thereafter. In this view pleasant dreams could have cumulative benefits on sleep, mood and well-being over time, in contrast to the downward spiral between nightmares and poor sleep and mental health.

So far, we have described how nightmares can be triggered by adverse events and how chronic nightmares can disrupt a natural function of dreams in adapting to adversity. While there is some consensus that this is the case, others and I suspect that nightmare sufferers are not defined by nightmares alone but are in fact 'highly sensitive' to all kinds of experience. That they experience the world as brighter, louder, more salient and more emotional. While this leads to being easily overstimulated, more susceptible to trauma, and even overwhelmed by dreams, nightmare sufferers are also creative, empathetic and more perceptually astute.

Those prone to nightmares also report vibrant inner worlds, including vivid and at times intensely positive dreams and daydreams. These dreams can fill dreamers with joy and wonder, like the following example: '*My family and I went to this amazing sightseeing place on the top of a massive mountain so when you looked out it looked like it was the edge of the earth. You could see the stars in the darkness above the clouds and the clouds were pink and orange and they were falling off the mountain like a waterfall. It was*

When Are Nightmares a Problem?

so stunning ... I felt happy when I woke up.' The positive, social and aesthetic dream brings a pleasant boost to mood in the morning.

Vivid and imaginative daydreams also feed into creativity, like the nightmare subject in our lab who reported: '*I briefly imagined small people exploring my organs ... then I imagined a ship sailing in a freshwater lake ... I could feel and smell the wind. It was peaceful.*' Given a few minutes alone with his thoughts, this nightmare subject's mind naturally wandered into a sensational and pleasant daydream.

Despite this upside to being nightmare-prone, prolonged periods of trauma or adversity clearly result in more severe nightmares, with worse consequences on multiple facets of waking health. Especially recurrent and post-traumatic nightmares can persist for decades without improving, and interfere with waking life in many different ways, such as being unable to stop thinking about nightmares or feeling distressed and irritable all day long. Nightmares also have immediate physical effects, a racing heart and gasping for breath, and cause frequent awakenings from sleep. These immediate effects are linked to poorer physical and psychological health over time.

Luckily, there are simple and effective treatments available for nightmares, which we'll take a look at next; these treatments will help those in the grips of nightmares to cultivate more pleasant dreams, counteract the negative effects of nightmares, and improve sleep and mental health overall.

Part III

Working with Dreams and Nightmares

5

Treating Nightmares and Going Lucid

I'm in a prison cell that has windows looking down into a dark and quiet kitchen. I'm looking through the window and I see two kitchen drawers opening by themselves, as if by a ghost. I have a feeling of dread, I'm really frightened, my hairs stand up on their ends and I feel all cold. I start crying and panic.

The nightmare recounted above was described by one of my research subjects. Though the specific details may be unfamiliar, the experience – the fear, a growing sense of things beyond one's control, the feeling of being overwhelmed – this will be recognisable to anyone who has ever had a nightmare. Within the dream, the intensity builds until suddenly, sometimes violently, the dreamer is thrust into awakening, filled with distress and agitation. The relentless and repetitive nature of nightmares makes one feel helpless, and when trapped in this cycle, many nightmare sufferers will resort to avoiding sleep and trying to erase any thoughts of the nightmare from their mind. Left untreated, the situation is untenable and unsustainable.

Before discovering which – and how – nightmare treatments work, let's remember what we've explored in the previous chapters: that nightmare disorder stems from recurring patterns of emotion dysregulation, both within the dream and on awakening. More specifically, our studies show that those with nightmares struggle to temper their negative thoughts and emotions – they perform worse on stressful tasks requiring cognitive control and display less activation in frontal brain regions, too, designed for dampening arousal when in distress. Nightmares flip dreams on their head, barring the benefits that are to be had from sleeping and dreaming well.

Thus, a key to recovery is to interrupt this vicious cycle. Nightmare treatments are designed to repair a process of adaptively working through stress and to promote more healthy emotional responses both in wakefulness and in sleep. In most nightmare therapies, the first step is to simply stop avoiding nightmares and to learn to face the unpleasant feelings associated with them instead. From there one can learn to actually transform the negative content of nightmares, to become a more active agent in dreaming, and change dream themes for good.

The nightmare I've just described follows the typical and all too familiar progression of nightmares: a negative image evokes fear, which builds in intensity and ends in panic. The dreamer feels unable to move or respond in any useful way. The first step of working with a nightmare like this can be to simply re-experience it in waking imagination, observing the feelings that are evoked with each scene. While this can be intense, the dreamer is reminded that they are in complete control of the visualisation and can pause it at any time.

Treating Nightmares and Going Lucid

In this case, the dreamer identifies the peak feelings in the final dream scene: she is terrified and alone. At this point, she is invited to flip the script, to change the dream in some way to assuage these unpleasant feelings. She says:

I think if there were other people in the room, I'd feel better. I could hold their hands, and I wouldn't be on my own. It just feels safer with other people.

As she imagines what this new dream would be like with her eyes closed, the visualisation takes on a life of its own and she reports:

We're standing there holding hands, and the drawers start going back in on their own. The room is stretching and getting longer and I can see more of the floor, dark and light checkered tiles, as the room stretches away. And the fear is going away. I feel calmer and lighter, and the dream has gone from grey to different shades of blue. I'm like a pale blue now. It feels quite powerful, like winning.

I have witnessed this time and again with my research subjects: as dreamers start to engage constructively with their nightmares in waking imagination, the dream responds in turn, spontaneously and creatively unfolding. When conscious attention is applied to the nightmare, it's as if a natural process of dreaming can continue, allowing the mind to overcome its sleeping fears in the space of waking imagination.

There are two major ways that this kind of nightmare

treatment can be achieved: from outside of sleep, by working with the dream in waking imagery (which then filters into sleep and helps to quash nightmares in their path); or from inside the dream itself, commonly known as lucid dreaming therapy.

Rewriting nightmares in waking imagery

The most common treatment for nightmares is imagery rehearsal therapy, which uses waking imagination to modify nightmares, as we've just seen. Imagery rehearsal therapy views nightmares as a sort of learned behaviour, where recurring themes replay a scripted (and unhealthy) response to a particular threat. The goal of therapy, then, is to rewrite the script and to practise, in waking imagery, a more favourable way of dreaming. These new patterns then emerge in the sleeping mind.

Imagery rehearsal therapy is based on two techniques: exposure and rescripting. First, the nightmare is recalled in the waking state, and patients are asked to re-experience the emotions and narrative of the nightmare. The goal of exposure is to habituate the dreamer, so they become accustomed to thinking about and feeling their bad dreams. In doing so, it lessens the fear that usually makes patients want to suppress nightmares.

In this first step, it's important to avoid feeling overwhelmed. This is critical because even while awake, recalling nightmares can trigger immediate stress reactions – a racing heart, sweating, panic. These are all learned responses to the nightmare's content. With imagery rehearsal therapy, one approach is to start with recounting more mild bad dreams, rather than a patient's most intense nightmare, and to gradually work up to re-experiencing more intense nightmares. In another variant of therapy, called 'exposure, relaxation and rescripting therapy', patients are taught several techniques of relaxation that help to counter the stress associated with nightmares. Relaxation can be especially helpful just prior to recalling a nightmare – most commonly progressive muscle relaxation, where the muscles throughout the body are sequentially tensed and relaxed, leading to a full body state of relaxation. Patients learn to observe how they feel and, only once relaxed, they begin to recall and retell the nightmare in detail. During the recounting of a nightmare, the dreamer continues to monitor their bodily state, and if signs of stress occur they can pause and return to relaxation, perhaps using the simple trick of slow diaphragmatic – or belly – breathing, in through the nose and down into the belly (whereas breathing shallowly into the upper chest is part of the body's stress response). These tools are useful during nightmare exposure, as patients learn to monitor fluctuating stress levels and return to a state of calm when needed.

In general, exposure is thought to support a process akin to fear extinction, which, as we explored in the previous chapters, is one proposed dream function that goes haywire in the case of nightmares. Thus, in part, therapy is an attempt to continue to accomplish the work of REM sleep, to lessen a negative response to a threatening memory or nightmare. When successful, the fear that accompanies a nightmare is gradually overridden by feeling unthreatened and calm in the face of this dream, and even other related bad dreams. The practice of exposure helps nightmare sufferers feel able to cope with and confront the unpleasant emotions that nightmares stir up, gradually lessening the distress caused by them. The ability to face nightmares head-on also counteracts one of the precipitating causes of nightmares, which is avoidance: patients usually go to great lengths to avoid thinking about nightmares, avoiding sleep for fear of having one. Unfortunately, avoiding nightmares has the unintended consequence of triggering their recurrence, like a dream rebound of suppressed thoughts. By removing avoidance, exposure dissolves this cause of nightmares, thereby lessening their frequency over time.

While exposure alone is effective in reducing both nightmare frequency and distress, it is usually only the first step in treatment. The second step involves imagery rescripting. In this exercise, patients are asked to come up with a new version of the nightmare, to alter elements or the ending of the nightmare script, as with the example of my research subject, who imagined other people in her dream to reduce feelings of isolation, and her dream transformed to a lighter feeling and shade. The change does not necessarily have to result in a positive dream, but it should feel satisfying or resolved in some way. To orient the dreamer towards where and how to rescript the dream, therapists will often draw attention to five common nightmare themes: safety – feeling in danger or unsafe; powerlessness – feeling out of control; intimacy – feeling a lack of closeness; trust – feeling unable to count on others; and esteem – not feeling good about oneself. Patients are guided to identify which of these themes are most prominent in their nightmare and to design a rescripted dream that overcomes this central threat. For example, a dreamer could find more power by taking control of the environment or other characters; if the dreamer feels unsafe they could construct body armour or turn down the volume or intensity of the dream. These changes can be added at any point in the nightmare, at the beginning, middle, or end. Once a rescripted version of the dream is written down, the instruction is to rehearse this new dream in imagination for ten to twenty minutes each day, ideally each night prior to sleep.

Over the course of therapy, successful rescripting leads to a progression of changes in dream content, as dreams move away from recurrent themes towards more original content. For PTSD patients, dreams become less like replays of a trauma; they start to include novel elements, weaving in new settings and characters even if the theme remains related to past trauma. In other words, the dreams start to become more creative and adaptive, more similar to healthy dreaming. In particular, dreams after trauma often feature prominent central images, like a tidal wave or impending attack, though as patients recover the images becomes less singular and intense, the dream pulling in more varied and nuanced content over time.

Importantly, patients who practise rescripting report a greater sense of control over their dreams, which is often called 'mastery' in nightmare treatments. This sense of control or mastery can be especially empowering for chronic nightmare sufferers or PTSD patients, who have long felt helpless against repetitive and potentially traumatic nightmares. What's more, mastery is strongly tied to positive treatment outcomes. When imagery rescripting is paired with nightly rehearsal, it reduces nightmare frequency and severity, improves sleep quality and lessens PTSD symptoms in patients with prior trauma. A sense of mastery is directly related to these improvements.

Besides instilling a sense of mastery, the rescripting component of treatment also leads to changes in the way PTSD patients think about themselves and about the world. Traumatised patients often develop what clinicians refer to as 'maladaptive beliefs' – unhealthy ways of thinking about other people, the world and themselves as a result of trauma. These beliefs often have to do with the nightmare themes discussed: power and control, trust, safety, intimacy and esteem. Patients feel they cannot control situations in life, or they see the world or other people as unsafe or dangerous; they do not trust anyone or want to be close to anyone, and even hold themselves guilty and at fault for what happened. By working through and rescripting these underlying themes in nightmares, patients develop more positive ways of thinking about themselves and the world in their waking lives. In other words, they develop more adaptive beliefs, and this corresponds with better treatment outcomes overall.

As alluded to already, these nightmare therapies are incredibly helpful beyond the scope of just reducing nightmares. In PTSD patients, nightmare therapy results in better sleep quality and less daytime PTSD symptoms (such as avoidance and hyperarousal). In fact, a recent study showed that after only a few sessions of exposure and rescripting therapy, 75 per cent of PTSD patients fell below diagnostic thresholds for PTSD (and these patients

Treating Nightmares and Going Lucid

had been suffering from symptoms for an average of eighteen years).[1] By treating nightmares directly, patients gain a better sleep quality and healthier sense of well-being, which supports long-term recovery from PTSD.

Given how simple and effective nightmare therapy is, some clinicians believe that nightmares should be the first step in overall treatment for PTSD. Nightmare therapies offer a gentle initial approach to helping people who have suffered a trauma, and for those who are wary of confronting trauma memories directly, imagery rescripting can be done with minimal exposure time and still yield benefits. To this end, some interventions have found that completing imagery rescripting *without* the exposure component of therapy can still be effective,[2] and we'll see later that this may be a preferred approach for some psychiatric patients. Even a simple intervention of rehearsing positive dream scenarios has been shown to benefit nightmare sufferers.[3] In support of this, when patients imagined a safe and comfortable dream with as much sensory and emotional detail as possible for ten to fifteen minutes a day for four weeks, this led to lower nightmare frequency and distress, with outcomes comparable to imagery rehearsal therapy.

More generally, it's worth mentioning that imagery rescripting can also be used to transform adverse memories (not just nightmares). In PTSD therapy, patients are sometimes instructed to reimagine their trauma memories with elements of rescripting, such as viewing the memory in slow motion or backwards, or as if projected in a movie theatre.[4] As an example, a victim of assault could rescript a memory to a version where they successfully defended themselves. Even though patients are aware that the rescripted memory is not accurate, experiencing these altered versions still helps to reduce the distress associated with the traumatic event.

Imagery rescripting can also be applied in the context of other psychiatric disorders such as social anxiety disorder and depression, where patients often have negative and intrusive waking imagery. To give some examples, patients with social anxiety disorder experience recurring intrusive images of being embarrassed or humiliated, often witnessing themselves from the outside. One patient repeatedly saw images of herself 'lurching around out of control, vocalising noises'; another imagined herself 'looking stupid with a red face'. These images are often based in real memories – the first patient remembers galloping around like a horse as a young child and being made fun of by her peers. In other cases, like eating disorders, patients have intrusive images of seeing themselves as overweight; one patient sees herself as if in a photo, with a fat stomach and thighs and double chin, eating chocolate and looking sticky and sweaty. The image evokes vivid sensory detail: the taste of sweets in her mouth, the smell of sweat, the feeling of tight clothes and a full stomach and hot, sticky skin.[5] Patients with anxiety disorders might repeatedly worry about past or future events and imagine worst-case scenarios. All of these are patterns of thinking that take the form of recurring, intrusive and uncontrollable negative imagery, like daymares. Similar to what happens in nightmare therapy, rescripting can help to modify harmful waking imagery, and this results in improved psychiatric symptoms and overall well-being. As a bonus, reducing worry or intrusive imagery in the period before bedtime will reduce nightmares for these patients, too.

To return to and summarise our waking nightmare therapies, these typically involve: exposure through writing and retelling a nightmare, and rescripting a dream and rehearsing this new dream each night. An additional component of relaxation can be added prior to exposure, so patients learn to monitor stress levels and return to relaxation as needed. In general, these treatments help patients become more able to cope with the unpleasant content of nightmares. They feel more in control of their dreams and are no longer afraid of or avoid nightmares.

In parallel to these cognitive and behavioural benefits of treatment, nightmare therapies also benefit patients physically, by reducing their levels of bodily arousal. Recall that nightmare sufferers are often in a state of hyperarousal – showing stress symptoms while both awake and asleep. These symptoms occur especially around bedtime, or on waking from or thinking about nightmares (like a racing heart or sweaty palms). Measures of physical arousal are in fact directly tied both to symptom severity and to treatment outcomes: in one study, nightmare sufferers' heart rate and frowning muscle activity (signs of stress in the body) were recorded during therapy sessions, and those patients who initially had stronger stress responses also had worse symptoms, including more disturbed sleep and greater nightmare distress. But successful therapy led to lower stress responses later on, and those patients whose arousal decreased the most improved the most, too. They had fewer and less severe nightmares, better sleep quality and fewer PTSD symptoms than patients whose stress levels remained high.[6]

In theory, nightmares are tied to physical threat responses that we're all familiar with from waking life: fight, flight, or freeze. Surveys have identified the most common nightmare themes to be: aggression or conflict (fight), being chased (flight), helplessness and illness or death (freeze). Whenever the mind thinks it is being threatened, the body reacts with these instinctual physical responses, and this is as true of the sleeping mind as the waking one. These are natural evolutionary responses to threat, but in the case of trauma and nightmares, they are triggered in contexts where they are no longer helpful. Thus, an underlying purpose of therapy is to help patients overwrite this unneeded threat response. In part, this is aided by practices of relaxation, teaching patients to remain calm when exposed to nightmare imagery. But relaxation is also helpful outside of exposure sessions, and can be purposefully applied elsewhere in life. For example, performing progressive muscle relaxation each night prior to sleep decreases presleep arousal, a known trigger for nightmares. And belly breathing can be used anytime during the day when stressful situations arise. In this way, patients learn how to recognise and regulate their stress symptoms across the day and night, thus lowering their baseline state of arousal (this can be especially helpful for traumatised patients, who are otherwise in a near-constant state of vigilance).

Beyond simple relaxation, nightmare therapy can help patients feel a physical sense of *safety* in relation to or even within their nightmare imagery. When we

Treating Nightmares and Going Lucid

feel a sense of safety, it overrides the fight-flight-freeze response, and within nightmares it opens up the possibility of other, potentially more empowering responses to a threat. One variation of nightmare treatment, called 'focusing-oriented dreamwork,' is especially helpful for finding a sense of safety within nightmares. 'Focusing'[7] is a specific technique of tuning into the body, while 'dreamwork' is just an umbrella term used to describe a number of approaches for working with dreams and nightmares in waking imagination. The purpose of focusing is to guide patients to access the physical sensations that occur in tandem with their emotions; as an example, alongside a general feeling of 'anger', the body might experience a squeezing heat in the chest, physical nuances that accompany this emotion. In the context of nightmare therapy, patients are instructed to first tune in to the physical feelings accompanying a nightmare and are then led towards finding a physical sense of safety, or 'finding the help', within the nightmare. These steps slot into an overall sequence of imagery work that bears a lot of similarity to the other rescripting therapies I've described.

The first step in this approach – called 'clearing a space' – has a similar purpose to relaxation. This step aims to clear the mind and ground the body at the start of a session and is as simple as taking a few minutes to get comfortable: take a few deep breaths, let go of wandering thoughts and become present in the body. In particular, patients are instructed to notice any distracting tensions or worries arising in the mind and gently put them aside; perhaps even visually imagine putting them on a shelf or in a box for the duration of the session. This allows patients to focus more wholly on dreamwork, observing whatever feelings arise in the body during the process. Throughout the session, dreamers are encouraged to come back to the grounded and self-observing stance that is established when clearing a space as needed.

In the second step, the dreamer is asked to recall and retell the dream or nightmare in detail, paying special attention to any bodily sensations that accompany the dream. If discussing the dream with a therapist or partner, questions can be asked of the dreamer to further clarify any sensations encountered in the dream and to invite the dreamer to gather an overall felt sense of the dream. The dreamer is asked to 'anchor' or 'provide a handle' for the overall feeling of the dream with a descriptive phrase, like this 'squeezing-hot-chest feeling of anger' or 'an electric jolt-in-the-stomach fear' or 'a gaping skeletal sense of loneliness'. This is similar to exposure, but asks patients to more fully explore the bodily experience of a nightmare (and as we learned, dreams are often fully embodied experiences).

The third step, then, aims to help the dreamer find a sense of safety *within* the nightmare. The dreamer is asked to 'find the help' in their dream, which means to look for any element of the dream that seems particularly energetic or potent, such as an unusually striking image or peculiar object or character, even if these images seem nightmarish on the surface. The dreamer is invited to embody or 'try on' these other elements of the dream and to explore what the dream feels like from this other perspective. This brings the dreamer beyond their usual first-person point of view of the nightmare (the view usually associated with strong feelings of fear and helplessness) and into an *other* point of view. This can bring an unexpected shift in the overall feeling of the dream and often helps patients to find a surprising sense of safety or support that is already available within the nightmare.

To give an example, one of my subjects was experiencing a bout of chaotic and fearful nightmares, including the following:

> I was trapped in a small room with five rabid tigers. I was trying to escape without letting any of them hurt me or get out ... I could hear the cats screeching and thrashing around the room in utter chaos.

During the nightmare, she felt powerless and terrorised by the big cats, but in retelling the dream, it was clear that the cats were the most energetic portion of the dream, what could be called the central image of the nightmare (as we learned about in Chapter 3). To experience the step of 'finding the help', the dreamer was asked to imagine *being* one of the cats, to try to feel what the dream would be like for them. Immediately she could sense their panic at being confined in a small room; the cats were desperate to escape. She realised they were not intending to threaten her; instead they felt scared and trapped, just like she did. Embodying the cats led to an immediate shift in this dreamer's understanding and overall sense of the dream and of what the dream needed to resolve.

This step of finding the help is then used as a segue into rescripting, where the dreamer is asked what the dream needs, or if the dream can continue in any way. In this case, after sensing the cats' panic, rather than focusing on her own escape (as in the original nightmare), the dreamer decided to open the door and let the cats run free. As she closed her eyes and imagined opening the door, to her surprise the cats simply lay down and became silent and still, no longer afraid. The dream had evolved and now evoked a sense of peace and calm instead of chaos.

Dreams often transform like this, in unexpected yet cooperative ways; a sort of improvisation emerges as the nightmare responds to the dreamer's attention. With focusing-oriented dreamwork, the goal is for the dream to continue organically and to allow a 'bodily felt sense of what is the right next step guide the dream forward'.[8] This differs from imagery rehearsal therapy, where the approach is more cognitive and patients are instructed to intentionally 'make up' a new dream. With focusing, the dream seems to arise more of its own volition as feelings are explored in the body, and especially those revealed when 'finding the help'.

To give another example of this, one of my subjects reported several bad dreams over the course of two months, all with a peculiar recurring theme of buses: buses speeding out of control, crashing in traffic pile-ups, or simply never reaching their destination. One morning he reported a dream of being frightened on a bus that was moving recklessly fast:

> Time in the dream sped by so fast that outside the sun was racing across the sky, and then the moon in turn, as if multiple days were passing while this bus barreled through the countryside.

Treating Nightmares and Going Lucid

The 'help' in this dream came from the landscape itself; the dreamer noticed that outside the bus he saw a beautiful scene – an empty desert filled with shifting light as each sun and moon raced across the sky. When he imagined being up in the sky, he saw to his surprise that the bus was hardly moving at all, inching along imperceptibly as each day and night passed in turn. By changing his perspective – observing from the sky above – an immediate shift occurred, in this case a sense of slowing down and appreciating the scene, which he felt as an opening in his chest and sinking down in his lower body. This subtle change made the dream feel resolved.

With focusing-oriented dreamwork, patients are often surprised by the help they find within their nightmare, the shift that comes from tuning in to what is already there.

In our studies we have found that the more subjects felt like they were able to relive their dreams during the sessions, the more positive their attitude towards dreams and nightmares over time; in other words, the exposure lessened their aversion to bad dreams. Subjects also developed an increased sense of mastery, of being able to change dreams when they were frightening. This correlated with a decrease in nightmare disturbances, such as trouble sleeping, physical stress from nightmares, or waking symptoms like anxiety or mood problems. Finally, it's worth mentioning that, while it might be tempting to try to interpret the dreams – to analyse what the cats or buses represent and how they relate to waking life – this is not at all necessary for nightmare treatment to work. The internal shift that occurs from engaging with the dream is enough.

Into the Dream Lab

To sum up, though the steps vary among different kinds of nightmare therapy, there are several clear patterns that emerge. Therapy is often applied to work with nightmares whose themes recur over time, as these represent learned threat responses. Patients learn how to better cope with their nightmares and develop a sense of mastery over nightmare content as they work with them in waking imagery. Especially when patients feel safe and relaxed, they can better recover from nightmares and nightmare distress. In general, these waking practices offer a means of transforming nightmares and instilling a sense of calm and control, which translates into more positive sleep and dream experiences in time.

While I've largely described how nightmare therapy can be aided through dialogue with a therapist, it can also be done on your own by following a few simple steps. In fact, multiple studies have shown that following nightmare therapies in self-help formats, such as in a book or online, are just as effective as in traditional therapy contexts.

If you have a nightmare to work with, you can try the steps of nightmare therapy yourself:

Step 1. Establish a sense of safety and relaxation. Clear your mind and let go of distractions, imagine temporarily setting aside troubles or worries; try progressive muscle relaxation or slow belly breathing to observe when your bodily state becomes calm.

Step 2. Recall and write down or tell (or draw) the

Treating Nightmares and Going Lucid

nightmare. Do this in present tense and first person and use as many sensory details as possible as you imagine the nightmare scene by scene. What are you seeing, hearing, feeling, thinking? Identify key emotions, bodily sensations and/or nightmare themes in the dream.

- During this step, continue to monitor stress and return to relaxation as needed: Try belly breathing, or work with a milder bad dream, or just the beginning of a nightmare, if it is too intense. Remember that this step alone can be therapeutic: learning to cope with the difficult feelings brought on by nightmares will lessen negative reactions in time.

Step 3. Rescript the nightmare. Once you feel familiar with the nightmare, you can work with the images in a new way. Try to 'find the help' and embody other elements in the nightmare; ask what the dream needs, find a bodily sense of the right next step, and allow the dream to continue. Or focus on repairing nightmare themes; come up with desired changes and write down a new version of the nightmare; try to be active, not avoidant, in changing the dream. Remember that what feels 'right' to you may not be obviously positive or triumphant. Focus on what would make the dream feel better and more satisfying when reimagined.

Step 4. Rehearse the new dream. Visualise this new dream for ten to twenty minutes each day in as vivid detail as you can, and anchor the safe, strong, or pleasant feelings that accompany this new dream.

In addition to these steps, nightmare therapy will be supported by good sleep hygiene: keep a consistent sleep schedule, relax before bed (don't exercise or eat a heavy meal) and practise relaxation and imagery rehearsal before sleep. Monitor and be aware of stress levels throughout the day, too, and practise simple relaxation techniques to regulate arousal. Signs of progress over time will include: fewer nights with nightmares, less severe nightmares and being less upset thinking about nightmares. Sleep quality will also improve, with less worry before sleep, less waking up in the night, and feeling better and more rested during the day. In sum, these simple steps for working with nightmares in waking imagery can be highly effective for treating nightmares, and can be easily implemented in therapy settings, in groups, or even by oneself.

Overcoming nightmares through lucid dreams

Now that we've established a basis for rescripting nightmares, we will focus on how rescripting can be applied from within the sleep state through lucid dreaming – where you are aware that you are dreaming while asleep. Lucid dreaming therapy can be added on to other therapies for nightmares, where in addition to waking imagery, patients learn – and it is a skill that can be learned – how to become lucid and to transform nightmares while asleep. Lucid dreaming therapy has enormous potential for treating nightmares and even PTSD, and the effects seen can be dramatic – a single successful lucid dream can lead to reduced nightmare frequency and distress and improve sleep as well as waking mental health. Experiencing a lucid dream can immediately re-form a patient's relationship to dreams, as they realise they have the power to change nightmares from within and to create positive and enjoyable dreams in their place.

To give some examples: in a recent study of fifty PTSD patients suffering from nightmares, patients attended a week-long online workshop where they learned techniques to induce lucid dreams (we'll go through these in detail later) and were instructed to generate dreams with a healing benefit once lucid. While awake, patients mentally rehearsed their dream plan, visualising the kinds of positive and healing dreams they wanted to elicit. They imagined becoming lucid and engaging curiously with their nightmares, perhaps seeking reconciliation rather than fighting or fleeing. Within just one week of training, 76 per cent of the patients had a fully lucid dream. In one account, the dreamer became lucid and called out for the dream to heal her, and then experienced her body 'vibrating with such force that she could hear roaring in her ears'. Another person requested to meet and befriend her anxiety, which resulted in the image of 'a giant, golden lozenge that evoked a feeling of amazement and gratitude'. These immensely positive dreams are polar opposites from usual post-traumatic nightmares and were just as impactful. After only one week of training, all of the patients had reduced nightmares and PTSD symptoms (even those who did not have a fully lucid dream), and all of them fell below diagnostic thresholds for PTSD. This outcome was maintained a month later, suggesting that far from being a quick fix, lucid dreaming can be a powerful tool with long-lasting effects in the treatment of nightmares and PTSD.[9]

In other studies, lucid dreaming has been added on to waking nightmare therapies. For example, in a nine-week group therapy for forty patients suffering weekly nightmares, in addition to sharing and rescripting nightmares, half of the patients were trained in techniques to induce lucid dreams. To give an example of how lucid dreaming facilitates nightmare therapy: one patient initially reported recurring nightmares of her ex-husband chasing her throughout the city in a car, and no matter where she ran, he would find her with a sadistic grin and nearly run her down before she awoke. Early in therapy the patient again had this dream, but she became lucid and yelled 'Stop!' to immediately wake herself up before the climax of the nightmare. Later in therapy the dream returned, but this time the patient became lucid and stayed in the dream, choosing to call for help, at which point other dream figures emerged and together they approached the man and he disappeared, resolving the nightmare. While all of the patients had fewer nightmares after therapy, those patients who achieved lucid dreams showed more immediate improvements in sleep and dream quality over the course of therapy. Those in the lucid dreaming group also appeared more motivated and enthusiastic, with an increased appreciation for their dreaming lives. Thus, while waking therapy alone effectively reduced nightmares, the addition of lucid dreaming boosted the benefits on sleep and dream quality.[10]

We have found similar outcomes in non-clinical studies, where having more lucid dreams was linked to better mood the next day – and in our case this benefit was seen even if the dreams were not *fully* lucid. To explain what I mean by that, it's important to know that lucidity is not an all-or-nothing phenomenon. Instead, levels of lucidity vary across a continuum. In semi-lucid dreams, for instance, we might start to notice that 'things in the dream seem too bizarre to be real life' and start to clue in to the fact that we are dreaming, without fully realising it.[11] This often happens in bad dreams, where, for instance, a dreamer might first feel afraid of falling off a towering ledge, but then some part of them has an inkling that there is no real danger. Even without being fully lucid, this partial awareness can be helpful to a dreamer and keep them from becoming too afraid. Along these lines, what we have found is that semi-lucid dreams are generally less anxious and fearful than non-lucid dreams and are associated with a more positive mood in the morning. Even a little bit of lucidity seems like a good thing. Of course, the largest benefits are to be had from fully lucid dreams, including a greater boost to morning mood and feeling more refreshed and energetic and even less stressed the next day.

In the context of nightmare therapy, becoming fully lucid has its clear advantages. In the first place, when someone is in the midst of a nightmare and realises they are dreaming, knowing without a doubt that the threat is not real brings a dramatic feeling of relief. This is similar to the relief one feels upon awakening from a nightmare and realising it was 'just a dream'. In fact, some lucid dreamers choose to wake up if they become lucid within a nightmare, just to stop the dream from developing any further, as in the following example of one of my subjects:

> I was learning to drive with my instructor, and I remember feeling frustrated because my feet could

> not move to press the pedals. At this point I realised I was dreaming because this is a recurring dream I have had before and recognised it. After realising it was a dream, I stopped the car and got out. The dream ended with my getting out of the car.

In this case, the dreamer becomes lucid upon noticing a typical bad dream scenario, recognising its recurring features. Once he becomes lucid, he decides to simply get out of the car rather than allowing the dream to further escalate, and then was able to wake up, not as a result of the nightmare climaxing but of his own volition.

Of course, lucid dreamers can also decide to stay in the dream and then use their lucidity to direct the dream more actively, rewriting the script from within. Many lucid dreamers first practise visualising alternate endings to a nightmare in waking imagination, to help them more easily tap into this alternate course of action once lucid. Indeed, most proponents of lucid dreaming therapy incorporate several techniques from waking dreamwork into treatment, such as having a clear intention of how to revise a nightmare ahead of time and rehearsing this desired dream before sleep to increase chances of success. In addition, there are several tips and tricks lucid dreamers can use to better control themselves, and the dream environment, once lucid, and we'll explore these in detail soon. Importantly, just as we saw earlier with waking imagery rescripting, even slightly changing the nightmare script, such as by getting out of the car or calling for help, can lead to effective resolution of nightmares.

That lucid dreaming provides both relief from

nightmare distress (when realising the threat is not real) and control over nightmare content should be familiar, as these are two of the keys to healing nightmares we discussed earlier. First, becoming lucid offers a reprieve from intense fear and overwhelm, as the dreamer has insight and awareness into the fact that what they are experiencing is just a dream. Second, becoming lucid helps to empower nightmare sufferers, giving them a means of control and a sense of mastery over their dream design. While waking rescripting therapies rely on learning these skills in waking life and having them translated into sleep, lucid dreaming has the advantage of tapping directly into the sleep state and applying these skills from within.

So how does lucid dreaming actually work to resolve nightmares? And what's going on in the sleeping brain while lucid? Let's revisit what we know about REM sleep, nightmares and emotion regulation, to see what lucid dreaming brings to the table.

Remember that one of the functions of REM sleep is to refresh connections between the brain's frontal regions and the amygdala each night, to support emotion regulation. When REM sleep is disrupted, as with nightmares, we become less effective at dampening arousal, both during sleep and while awake. Within nightmares, patients feel helpless and out of control, unable to manage or respond to threats, and are often physically overwhelmed into awakening. This leads to a vicious cycle: nightmares cause waking distress and weaken our defences against stress the next day, leading to further nightmares and disturbed sleep the following night, and so on.

Lucid dreaming seems to show the opposite patterns.

For instance, lucid dreamers display heightened cognitive skills within dreams, being aware of and in control of their mind, body and feelings (for example, lucid dreamers report being able to make decisions and control their actions in dreams). These are useful tools for emotion regulation, observing how we feel and deciding how to respond, rather than feeling helpless, avoidant, or overwhelmed. Even in waking life, frequent lucid dreamers report a greater sense of control over their life – a higher 'internal locus of control', meaning they feel that they are in charge of their actions and decisions and outcomes in life. People who practise mindfulness and meditation also have more frequent lucid dreams, likely because they are more self-aware and mentally in control in waking life, which spills over into more lucid dreaming.

These findings converge onto the idea that lucid dreaming is linked to better cognitive control in both wakefulness and sleep. Some dream scientists, myself included, think this means that lucid dreaming can actually help with regulating emotions during REM sleep. If we remember the process model of emotion regulation described earlier, we can see how lucid dreamers are able to control their attention and their thoughts and behaviours in response to stressful scenarios in dreams; they can divert their attention from threats or focus on calming their emotions, and they can change the dream narrative or environment to make it more enjoyable. In the example given earlier, the lucid dreamer first simply yelled 'Stop!' to end a nightmare altogether, and later recruited the help of other characters to confront and overcome an attacker. These are proactive ways to handle a stressful situation.

Treating Nightmares and Going Lucid

Lucidity gives the dreamer agency and a sense of control, tools they can use to manage their emotions, just as we do in waking life. In fact, lucid dreams are also associated with increased frontal brain activity, especially in the prefrontal cortex, during sleep. This could be in part what enables these higher-level cognitive skills within dreaming and could also support emotion regulation during the night. While it's too soon to say if this is how lucid dreaming works, it is the focus of current research in my lab and others', to find out whether and how lucid dreaming benefits the emotional functions of sleep.

While there's clearly much more to learn about the neuroscience of lucid dreaming, the important thing so far is to know that lucid dreaming therapy works – it improves sleep and dream quality in patients with nightmares and offers a fun and empowering way to rewrite nightmares from within.

With all that in mind, let's turn to the biggest hurdle of lucid dreaming therapy: learning how to become lucid in the first place.

Fifty years of lucid dreaming research so far confirms that most of us have had at least one lucid dream, though the majority of us have lucid dreams only rarely (close to 80 per cent of people have them less than once a month). Luckily, lucid dreaming is a learnable skill, and there are several simple and reliable ways to increase your chances of becoming lucid each night.[12]

The first step to lucid dreaming is to become familiar with typical dream patterns, most often by keeping a dream journal. Over time, a dream journal can reveal the presence of peculiar recurring dream themes, such as being late to an exam or having a dead relative cropping up. These recurring features can serve as what's called 'dream signs' – repeating bizarre or impossible dream themes that can clue us into the fact that we are dreaming ('oh, my clock is not working again, hang on, I must be dreaming!'). We met an example of this earlier, when one of my subjects reported: *I realised I was dreaming because this is a recurring dream I have had before and recognised it.* He noticed the recurring theme of driving in a car with broken pedals, and so realised he was dreaming.

Since recurring themes are so common for those with nightmares, they can be excellent clues to pay attention to while dreaming. One of my subjects became lucid twice in a single night because she noticed a similar bad dream theme repeating. In the first dream, she and a group of friends were going surfing when an argument broke out on the beach. The sudden intensity of the argument in an otherwise pleasant dream caught her attention, and when she realised it was a dream, she decided to wake up. Later the same night, she dreamed of being at a birthday party when a bunch of women started arguing aggressively. She became lucid by recognising that the situation was similar to what she had experienced before, in her previous dream, and this time decided to stay in the dream and mediate the dispute. This is a fairly subtle recurring feature, a sudden intense argument, but it goes to show that by paying attention to dreams, nuanced patterns can emerge, repeating topics or emotions or body sensations, like a sudden argument or a speeding car.

As mentioned, the best way to uncover recurring dream

Treating Nightmares and Going Lucid

signs is to keep a dream diary, to start to notice dream patterns and later recognise them while asleep. Dream signs can also be linked to another technique termed 'reality testing', where for a few moments several times a day – thirty seconds or so is enough – you ask yourself whether what you are experiencing is like a dream. This can be done especially when something unusual or bizarre happens during the day, as this trains the mind to perform reality tests whenever something dreamlike happens, which is of course bound to happen within a dream. As a waking practice, reality tests are similar to a moment of mindfulness, stopping to really observe and pay attention to the quality of one's experience. After all, mindfulness and meditation practices elevate how aware we are not only during the day but also when asleep and dreaming. Unique from mindfulness, however, is the instruction to further 'test' reality by doing something that would differ in a dream, like looking at the time on a clock or plugging your nose and trying to breathe. In a dream the numbers on the clock will likely be unstable, and you will probably be able to breathe through your nose even when it is plugged. By doing reality tests repeatedly during the day, especially when something dreamlike happens (or when a dream sign occurs), this habit will carry into dreams and lead to the realisation that you are dreaming.

While dream journaling and reality tests are practices done during the day, another more physical means of inducing lucid dreams is through intentional sleep disruption. Lucid dreaming most often occurs during REM sleep in the morning (when dreams are altogether at their most vivid and intense). Because of this, one of the most successful techniques for inducing a lucid dream is to wake up in the morning, about an hour or two earlier than normal, and stay awake for five to twenty minutes before going back to bed. This is called the 'wake-back-to-bed method', and it was developed by lucid dream pioneer Stephen LaBerge. Because of the structure of sleep cycles as we saw earlier, our mornings are mostly filled with REM sleep, and falling quickly into this stage increases chances of achieving a lucid dream. Even using the snooze button on an alarm clock is associated with more lucid dreams; in one study, those who hit the snooze twice or more had the most lucid dreams, likely because they were waking from and quickly re-entering REM sleep in between buzzes.

To supplement the wake-back-to-bed method, the short window of wakefulness in the morning can be used for mental rehearsal, such as setting a strong intention to become lucid or even visualising a lucid dream. Dreamers can repeat a phase in mind before drifting off to sleep like 'The next time I am dreaming, I will remember that I am dreaming', creating a form of prospective memory, that is, a memory to do something in the future. We do this all the time in our daily lives, a commonplace example being something like: 'On my way home from work I will remember to buy milk.' The difference here is that the goal is to *remember to become lucid* later when one is dreaming, which means that a person first has to change states of consciousness and then recall this prospective memory once asleep and dreaming, precisely when memories are most difficult to access. Despite all these hurdles, setting an intention like this during wake-back-to-bed is surprisingly effective: Laboratory studies show that 25 per cent of subjects are successful on their first night of doing this combined practice, and this increases to 50 per cent on a second night.[13]

More generally, combining all the techniques we've covered gives the best chance of success: complete several reality tests during the day, schedule an early morning awakening, and set the strong intention to become lucid before falling back to sleep. The morning wake period can also be used to recall a dream from the night: write it down and look for any immediate dream signs (bizarre elements that could clue you into the fact that you are dreaming), then set the intention to become lucid if these dream signs reoccur. This is helpful because dream themes tend to reoccur across a given night, so dream signs from an early awakening are more likely to reappear in a subsequent sleep period. When used in combination, these techniques increase lucid dream frequency by over 85 per cent in a week, even with only a five-minute awakening in the morning and minimal disruption to sleep.

While these techniques are useful to any lucid dreaming enthusiast, in the context of nightmare therapy there is one other lucid dream skill that is highly relevant: learning how to *control* lucid dreams. In many cases, when people become lucid for the first time in a nightmare, their first thought is to simply wake up, as it is an easy and immediate solution to end a bad dream. But choosing to stay in the dream opens up the possibility of creating more positive and transformative dream experiences in the place of nightmares, which can be more impactful in the long run. Of course, the level of control available in lucid dreams varies widely: sometimes it may be hard just to maintain lucidity, whereas other times lucid dreamers are able to exert a high level of control over themselves and over the dream, like flying or walking through walls or making people or objects appear on command. Luckily, controlling lucid dreams is also a learnable skill.

The first step to controlling dreams is to focus on self-control, which means being able to control one's own thoughts and emotions and body in the dream. Beyond simply being aware of the dream state, dreamers start to interact with the dream intentionally, consciously choosing what to think, where to move, and how to respond to dream content. As an example, if a dreamer becomes lucid within a nightmare, they could first manage their emotions by choosing to remain calm and practising techniques to relax, such as breathing slowly and deeply. The dreamer can then even generate positive feelings within the dream, like joy, curiosity and optimism, even if the dream seems threatening. Approaching the dream with this positive attitude will often lead to favourable shifts in the imagery, similar to the examples given earlier with waking rescripting. The first time I became lucid in a nightmare, I was being chased by a monster and when I realised I was dreaming, I stopped running and turned to face the monster calmly and curiously, asking, 'Who are you?' It instantly transformed into a friend of mine, someone I'd had an argument with recently. Simply becoming lucid and changing my response – my emotions and actions – led to a revelation in the dream that made it no longer fearsome.

From this foundation of self-control, lucid dreamers can then turn to several strategies to gain control over the dream world, to manipulate the dream environment or

Treating Nightmares and Going Lucid

transform other characters at will. These strategies often build on the simpler forms of self-control we've just seen, like choosing where to look or how to act in a dream.[14] For example, lucid dreamers can use their voice to influence the dream, telling another character, 'You are my guide!' or shouting out commands like 'Pause!' or 'Disappear!' Verbal commands can also be combined with actions in the dream environment, such as stating an expectation – 'through this door I will find Paris' – and then walking through the door; or wondering aloud what you will find in your pocket as you reach in for an object. Expectation plays a big role here – most lucid dreamers report that they have to believe and expect the dream to change in order to be successful. Dream bodies can also be used in creative ways to up the ante on dream control. Flying – one of the most popular dream activities – sometimes requires a bit of extra effort in a lucid dream, such as flapping your arms repeatedly like a bird, raising your fist in the air like Superman, or jumping higher and higher off the ground. Dreamers can also play with their attention to manifest objects in the dream. In a study where lucid dreamers re-created a laboratory scene in their dream, one subject reported: *'I would close my eyes, think of an object, open my eyes, and it would appear.'*[15] With practice, lucid dreamers can learn to populate their dream world with desired objects and people, and even mould the setting itself to their command. But even these advanced forms of dream control usually build on simple tricks, like using one's speech or actions and strong expectations to manipulate the dream at will.

Into the Dream Lab

If you want to learn how to induce and control lucid dreams on your own, here are the tried and tested techniques you can use:

1. **Dream signs.** The first step for aspiring lucid dreamers is to keep a dream journal and to become aware of recurring dream themes, especially any unusual or peculiar dream elements that could clue you in to the fact that you are dreaming. Learning to recognise these dream signs will help you to become lucid while dreaming.
2. **Reality testing.** This technique trains your mind to be more lucid during the day. Several times a day take a few moments to carefully observe your experience; to 'test' reality, try to push your finger through your hand, turn on a light switch, or read text in a book (all of which typically fail in dreams) to establish whether you are awake or dreaming.
3. **Wake-back-to-bed.** Temporarily adjust your sleep patterns to try to induce lucid dreams. Wake up a couple of hours earlier than usual, stay awake for about ten minutes, and then return to sleep to increase chances of lucid dreaming (remember that REM sleep is more abundant in the morning, and lucid dreams are most common in morning REM sleep).
4. **Intention setting.** Practise some form of visualisation or mental rehearsal to prepare the mind just prior to a morning sleep period. This might include repeating a phrase like 'The next time I am dreaming, I will remember that I am dreaming.' Or imagine yourself in a recent dream, and notice a dream sign that could help you become lucid.
5. **Dream control.** While all of these techniques have the

Treating Nightmares and Going Lucid

goal of inducing lucid dreams, remember that levels of control vary across a spectrum. Sometimes, you might have only a little agency in the dream, controlling your own actions or emotions, but you can build on this self-control – try using your voice or body to shape dream content to your will.

You have about a 50 per cent chance of having a lucid dream within a week using a combination of these techniques. With practice, lucid dreaming can help nightmare sufferers, and anyone, to experience immensely positive and healing dreams in the place of nightmares.

One thing to keep in mind is that even within fully lucid dreams, you will not obtain absolute control over dream content; instead, dreams unfold in a manner similar to the waking rescripting I described earlier, where the dream retains some spontaneous creativity in response to the dreamer's direction. As an example, a dreamer might summon another character, but the character responds in unexpected ways, acting of its own volition; or a desired scene of Paris might start to look more like a cartoon than the real thing. In an experimental study where lucid dreamers tried to re-create a laboratory scene in their dream, they were often successful at manifesting objects (like a desk, a rubber snake, some plastic fruit and a clock set to six fifteen), but the dream objects then took on a life of their own, as one subject reported, 'The clock spun to midnight ... the snake moved off the metal desk and wrapped around the fruit.'[16] This is typical in lucid dreaming, and especially in the context of nightmare therapy it's important for the dreamer to keep an open attitude towards the dream rather than resisting or becoming tense or frustrated with slippery dream control, which could fuel a more negative dream. A certain cognitive flexibility is required – both directing the dream and allowing it to unfold of its own volition at the same time.

This is part of the wonder of both waking dreamwork and lucid dreaming: learning how to co-create healing dreams in cooperation with the dreaming mind, no longer at the mercy of nightmares.

By visualising alternate endings to nightmares during the day or training the waking mind to become more lucid, the healthy terrain of imagination becomes a resource for the dreaming mind. Through both waking therapies and lucid dreaming, nightmare sufferers learn to respond with more agency and to find a sense of calm and safety in the face of nightmares. These simple practices help patients to better cope with nightmare emotions and empower longtime sufferers to change the content of dreams for the better. In the end, the benefits of nightmare therapy extend beyond relieving nightmares; they improve the overall quality of dreams, restore sleep and support waking mental health as well.

6

Engineering Dreams

In the prior chapter we learned about the main therapies for treating nightmares, which rely on waking imagination and lucid dreaming to modify the content of dreams. In a way, both of these therapies attempt to carry intentions from waking life into sleep in order to diminish the frequency and intensity of nightmares. In this chapter we will take a step farther, into the world of dream engineering, and explore technologies that aim to elevate our influence over dreaming even as it is occurring during sleep.

The tools we use for engineering dreams are designed to modulate dreams by stimulating the sleeping brain and body. Dreaming is, after all, an embodied experience; dreams unfold in our minds, brains and bodies in real time, and sensations from the real world often seep into this simulated world. Because of this, dream engineers like myself have learned how to use sensory input to direct the content of dreams. I like to think of dream engineering as similar to the ambient elements of a cinematic production, how the lighting, the musical score, the pace and the action guide the meaning of a film. Of course, dreaming is not nearly as predefined as a cinematic script, and dream engineers cannot write a predetermined narrative for the dreamer to follow. No, the scaffolds of the dream world continue to be filled in with the memories, thoughts and emotions of the dreamer, beyond the control of an engineer. But we *can* nudge some of the parameters of dream design towards desired content.

What are some of the ways scientists use to engineer dreams? And what happens when we introduce outside elements to the dream world? To what extent can these stimuli direct dreams?

Just as the dreaming brain can react to stimuli like the noise of a door slamming, and responsively create a dream to match that stimulus, so, too, can it react to any number of stimuli we might want to use to impact dreams. Some of these elements include simple lights, sounds, or physical sensations like touch, temperature, or vibration. When these stimuli are presented to a sleeping person, the resulting sensations can influence dreams through implicit associations in memory – a pleasant scent to induce more positive dreams, or limb pressure to elicit movement in dreams.

While in the last few years science has shown that the sleeping brain is processing much more sensory information than previously believed, it's important to be conscious of how far you can take this without interrupting the sleeping state. Of course, selected stimuli cannot be so arousing that they cause an awakening or disturb sleep. Thus, dream engineers are limited in how much sensory input they can transmit to sleepers from the real world; they are restricted to relatively low intensity and periodic stimuli, having small though measurable effects on dreams. And, since the content of dreams is idiosyncratic, that is, based in our own memory, experiences and concerns, this means that as sensations seep into the dream world, the mind makes sense of them through personal references in memory. The scent of lilac reminds me of spring's bloom, or the light pressure on my leg reminds me of my cat brushing by. These subtle sensations filter into the dream world unobtrusively, all while a person remains asleep and amid a dream.

A variation of this approach, called 'entrainment', uses sensory cues to mechanically manipulate the rhythms of sleep, such as repetitive clicking sounds or puffs of pressurised air. These pulsing stimuli push and pull at the rhythmic beating of the sleeping body – like the oscillations of firing neurons, the cyclic in-and-out of breathing – all of these physiological tempos and cadences that affect the mood and manner of dreaming. Even synchronised electrical currents applied to the scalp can amplify or inhibit brain waves. These approaches can guide general attributes of dreaming, augmenting awareness or suppressing recall, lightening emotional tone or injecting some movement into dreams, to give some examples.

While the approaches above are applied only within sleep, a more targeted way to steer dream content is by combining stimulation with some of the more traditional therapies, such as adding a component of exposure or training prior to sleep. This is the basis of a method called 'targeted reactivation', where sensory cues are first presented in the presleep window alongside a task, and the same sensory cues are then re-presented after a person has fallen asleep, to *reactivate* the associated mental content.

As an example, subjects might practise a vocabulary learning task prior to sleep while at the same time they are exposed to stimuli such as scents or beeping sounds. This somewhat Pavlovian trick trains the brain to associate the memory of the task with the sensory stimulus. Later, when the subject is asleep, the same stimulus is presented – the beeping or odour – and this causes the task-related memories to be reactivated (and strengthened, thanks to a natural function of sleep in memory consolidation). While sleep scientists use targeted reactivation most often to enhance learning, which we'll explore later, dream scientists can use targeted reactivation to influence dreams themselves. Especially of interest in nightmare therapy, targeted reactivation can be combined with waking imagery rehearsal, or even lucid dream training, to increase the success of these therapies within sleep.

In the last chapter, we saw how nightmare therapies typically rely on some continuation or reappearance of a waking practice into dreaming – this might include visualising a positive dream during the day or setting the intention to have a lucid dream. Dream engineers capitalise on a more immediate access point to dreaming, the brain-body interface, to stimulate and shape dreams during sleep. Dream engineering, then, can act as a bridge for waking interventions to find their way into sleep, and a primary goal of my work is to develop these tools to enhance existing nightmare therapies, to provide support to patients while asleep and dreaming, and to target and diminish nightmares when they occur. Dream engineering used in this way can offer practical benefits to sleep and mental health, as we'll see throughout this chapter.

Engineering lucid, pleasant dreams

Targeted reactivation is, so far, the best conduit to channel waking imagery into a night of sleep and to enhance traditional nightmare treatments. In my own work, we use targeted reactivation to induce lucid dreams, with an eye towards enhancing lucid dream therapy in the future.

I first led a lucid dream engineering study in my earliest job out of graduate school, working in a sleep lab set beside the sprawling Swansea Bay in Wales. The study took place in the early morning – at 7:30 a.m. – so subjects had to wake up very early to come to the lab. This created a kind of wake-back-to-bed commute, where an early morning awakening at home was followed by a quick trip to the lab and a return to sleep shortly thereafter. This seemed easier than spending the whole night in the lab, though I vividly remember racing my bike along the dark and windy shores of Swansea Bay each morning, anxious to get to the lab on time. I spent those early mornings monitoring subjects quietly from the control room with my close friend and fellow researcher, Karen Konkoly, patiently awaiting the characteristic eye movements that signalled the onset of REM sleep.

I'll never forget the first days of the study, watching as one of our subjects slipped into sleep, and gradually the muscles of her face relaxed and any traces of muscle tension on the EMG disappeared; her body was entering paralysis, indicating that she should be dreaming soon. Then her eyes started darting: REM sleep had begun. 'Play the cue!' I whispered eagerly, and Karen flashed the LED light in the bedroom on and off three times. We stared at the screen in excitement, but the subject just kept sleeping, seemingly undisturbed. Half a minute later Karen triggered another cue, this time a series of three beeps presented through speakers by the bed. Then it happened, a sudden spike on the eye channels – once, twice, three times quickly. This was the signal. Though the subject was still asleep in dreamland, she had gained control of her eyes and was speaking to us in left-right glances. 'I can hear you!,' she said. 'I'm lucid.'

We both gasped and pointed at the monitor, grinning and trying not to make any noise. We continued the protocol, flashing the lights and waiting for a response, playing the beeps and watching for a signal. The subject kept responding in turn, saying, 'I got your message! I'm lucid.'

Although we had no way of knowing what she was dreaming at the time, she later told us she had been eating dinner with her family at the kitchen table when the overhead lights began to flicker. She didn't understand for a moment what was happening, absorbed in a gloomy conversation about a recent death in the family. When she heard the beeping sound she remembered, 'Oh, this is a dream!' She moved her eyes as instructed and, realising it was a dream, she decided to change the conversation to something more convivial, creating a pleasant family scene. She kept dreaming while watching for the flickering lights and listening for the sound cues, which were seeping into her now lucid dream.

This is the crux of dream engineering: though in sleep we may appear to be cut off from the external world, there continues to be a flow of input from our senses, which is often seamlessly incorporated into a dream.

To increase the chance that our light and sound cues

would be recognised in the dream, we trained subjects before sleep to pay attention to these cues and to associate them with a lucid mind state, a form of 'targeted reactivation'. In this case, we asked subjects to lie in bed awake but with their eyes closed for twenty minutes prior to their morning nap, and we presented light and sound cues at one-minute intervals, with verbal reminders to observe the cues and to 'remember to become lucid' each time they noticed a cue. When subjects later entered REM sleep, we replayed these same cues, hoping subjects would observe the cues in their dreams and, indeed, remember to become lucid. In just one morning nap, half of our subjects had fully lucid dreams complete with left-right eye signals, and a few subjects even had their first-ever lucid dream, right there, in the lab.[1]

This study was a success largely because of how efficient it was, inducing lucid dreams, however briefly, in a single morning session. But the idea to use sensory cues for inducing lucid dreams came around much earlier, especially in studies by Stephen LaBerge, who developed the DreamLight in the late 1980s – a sleep mask that could detect eye movements using an infrared sensor in the mask, and upon detecting REM sleep presented flashing red lights to the sleeper. When used in combination with both wake-back-to-bed and some presleep mental rehearsal, over 80 per cent of subjects using the Dream-Light experienced at least one lucid dream in a month. In a classic example one subject reports: '*I was walking along a road with my boss and the whole scene flashed, cueing me that I was dreaming. I mentioned it to him, and flew a little to prove it.*'[2] In general, preparing the mind seems just as important, if not more so, than presenting sensory cues. When presented only with cues, in the absence of mental preparation, 20 per cent of research subjects at most will become lucid in a single night. That number can leap to upward of 80 per cent when subjects practise reality testing and intention setting in the days to months prior to cued induction attempts.[3]

Besides the contribution of mental preparation, certain individuals may be naturally more sensitive to sensory cues and more likely to notice and incorporate them into their dreams. In fact, one important finding from our morning nap study was that nightmare sufferers observed the sensory cues in their dreams more often than those without nightmares (and 60 per cent of subjects with nightmares became lucid, compared to 40 per cent of those without nightmares). This aligns with our theory that nightmare sufferers are highly sensitive – more attuned to perceptual stimuli while awake – and suggests they are also highly sensitive to perceptual stimuli while asleep and dreaming. This is especially important and helpful if we want to use sensory cues to engineer lucid dreams in the context of nightmare therapy.

Nevertheless, one important question for dream engineering to tackle is: how much sensory stimuli can get into your dreams? And how can we better train people to notice sensory cues while dreaming?

In my Dream Engineering Lab in Montreal, we looked to answer some of these questions in a recent study with scientists in the Netherlands and Italy. We followed the same basic protocol of pairing presleep training with sensory cues and replaying the cues during REM sleep to trigger lucid dreams. But to increase the chance that subjects would perceive the cues while dreaming, we adjusted the presleep training to one of my favourite lucid dreaming techniques: the senses initiated lucid dreaming technique.[4] Rather than just repeating an intention to become lucid, the technique involves carefully observing one's sensory experiences just before sleep, paying attention to visual, auditory and bodily sensations in turn.

In our protocol, subjects lay in bed with their eyes closed and followed instructions to pay attention in sequence to any visual images arising behind their closed eyelids, any sounds occurring in their environment, and any physical sensations in their body. We combined this sequence of sensory observation with real sensory cueing: when the subject was focusing on their visual experiences, we presented a flashing red light; when they focused on their hearing, we presented a beeping sound; and vibration cues were presented during the bodily attention segment. The expectation was that this cued presleep training period would increase subjects' perception of these cues later during REM sleep. Overall, again around 40 per cent of our subjects became lucid in a single morning nap when the cues were re-presented in REM sleep, replicating our prior work. Moreover, almost 30 per cent became lucid even if they were *not* cued during REM sleep, suggesting that even alone, this sensory-based presleep training is fairly effective.

You can try the senses initiated lucid dream technique yourself at home. The goal is to train your mind to pay attention to visual, auditory and bodily sensations in turn, while lying in bed prior to an early morning sleep period, in the following steps:

Step 1: Focus on sight: Your eyes should be closed and relaxed and relatively still – simply pay attention to the darkness behind your closed eyelids.

Step 2: Focus on hearing: Shift your attention to your ears and listen for any subtle sounds in your surroundings; if it is quiet, you might hear the sound of your breathing or heartbeat.

Step 3: Focus on bodily sensation: Direct your attention to your body and notice any sensations such as tingling, heaviness, lightness and so on. You might also feel the weight of the blanket or the temperature of the room.

Practise cycling through these senses quickly at first, spending a few seconds on sight, then sound, then body sensations, a handful of times. Once you get the hang of it, then slow down and focus for about thirty seconds on each step. Go through four slow cycles, which might take about five to ten minutes in total to complete. As time goes on, you will probably start to feel relaxed and you might even witness some hypnagogic imagery sprouting up, such as seeing lights and images, hearing noises or music, or feeling sensations of falling or floating. This just means that you are starting to fall asleep. Continue to pay attention to each new image or sensation without trying to control them in any way. If you get distracted or your mind wanders, simply come back to begin a new cycle. After you have finished the four cycles, you can stop and fall asleep, more likely to have a lucid dream.

In some instances, people report that this technique leads to what's called 'wake-induced lucid dreams', where they maintain awareness as they fall asleep and immediately enter a lucid dream. Others report vivid false awakenings, a phenomenon where a person dreams that they are waking up in bed, when really they are in a dream. In some cases, the dreamer might be practising the sensory technique and then suddenly find themselves in a clear visual scene of their bedroom. It feels like they are awake, as the dream recreates their bedroom scene with almost no sense of shifting into sleep. If the dreamer can *remember* that they were in fact just practising the technique with their eyes closed, they can realise that this is just a dream, and they can get out of bed and enter the dream fully lucid.

This brings us to another concept called 'dream incubation', where visualising or reflecting on something just prior to sleep can lead to dreams with related content, a sort of carry-over where the experiences, thoughts and intentions of the waking mind flow into dreams. This can be useful in nightmare therapy, to carry pleasant and enjoyable themes from the presleep state into dreams.

On its own, dream incubation seems to work about half of the time, with judges rating 45 per cent of subjects' dreams as related to a personal problem intentionally mulled over just prior to sleep.[5] Dream engineering can dramatically increase these odds, with over 90 per cent of subjects rating dreams as related to auditory stimuli that were presented in the liminal state as they fell asleep.

This method is called 'targeted dream incubation' and borrows principles from targeted reactivation but pairs them with dream incubation, linking auditory cues to mental rehearsal in the sleep-onset state, on the borders of dreamland. The Dormio device, one piece of technology designed with targeted dream incubation in mind, is a glove-like wearable that uses sensors to detect when the wearer shifts into sleep based on changes in heart rate, electrical activity on the surface of the skin, and muscle tension in the fingers.[6] As the wearer falls asleep, Dormio delivers a spoken prompt to incubate specific content into dreams. The prompt could be anything, specific to the dreamer and depending on what they would like to dream about, such as 'remember to dream of flying' or 'dream of your favourite childhood memory'. A couple of minutes later, after the glove senses that the wearer has fallen asleep and entered dreamland, another prompt awakens them: 'Can you tell me what you were dreaming about?'

The process of falling into sleep with a prompt and being awakened to report a dream is repeated several times in a row, so multiple brief dreams in the range of seconds to minutes are collected. As this process goes on, the sleep onset images become progressively more dreamlike and creative, the brain seemingly entering a more pronounced sleep with each dip. In one study, subjects were reminded to simply 'remember to think of a tree'. The dreams at first were more akin to thoughts, of 'trees, many different kinds, pines, oaks' and later evolved into more dreamlike dreams such as 'a tree from my childhood, from my backyard. It never asked for anything'. After five sleep onsets in a row over the course of about forty-five minutes this same subject finally dreamed: 'I'm in the desert, there is a shaman sitting under the tree with me, he tells me to go to South America ...' Over the course of several studies, at least 90 per cent of subjects were able to fall asleep holding the desired content in mind, generating multiple creative microdreams related to a given prompt.

While there are many uses of this technique that we will explore, let's start with its potential value in treating nightmares.

If you remember, one important factor in how well patients respond to nightmare therapy is their belief in their own ability to control a nightmare's content. For therapy to work, patients need to move away from feeling helpless against nightmares, towards realising that they *can* control their dreams. This is often called mastery, but it is more broadly a form of self-efficacy – one's belief in their own ability to succeed. In general, a patient's belief in whether they can benefit from therapy and recover is a strong predictor both in how much effort they will put into therapy and how well they will respond. Importantly, clinical research from other fields shows that boosting self-efficacy either prior to or alongside psychological treatments can lead to better outcomes. By contrast, a lack of self-efficacy limits how well any type of psychotherapy will work, and some patients simply do not partake in or do not adhere to therapy because they do not believe it will work. This is especially the case for nightmare therapy, where patients with recurring nightmares or PTSD may have had the same nightmare playing on repeat for decades, and it seems ludicrous to them to think they could simply redesign these haunting dreams.

In this context, a practical precursor to nightmare therapy would be to augment a patient's belief in their ability to control dream content – in other words, to increase their self-efficacy and thereby improve their response to therapy. Arguably the easiest way of doing this is to provide patients with clear and concrete experiences where they are able to control their dreams. Using Dormio is one way to give nightmare sufferers quick and easy evidence that it is possible to direct dream content. We have found that a single session of targeted dream incubation elevates subjects' perceived dream control and self-efficacy and we think this could be helpful to introduce to patients just prior to initiating therapy, or in multiple sessions alongside therapy. As patients begin to feel some control over their dreams, they may well be encouraged to embark on or continue with rescripting therapy, improving the likelihood of recovery.

In addition to providing simple experiences of dream control, targeted dream incubation can also be used to generate healing or positive imagery, little capsules of pleasant dreams.

In a way, sleep onset microdreams are more accessible than dreams later in a sleep period and are easier to induce yet similar to lucid dreams, where some conscious attention is applied to shape dreams. With this in mind, I worked with lucid dreaming expert and instructor Charlie Morley,[7] who developed a twenty-five-minute guided meditation designed to bring someone into a sleep onset state and induce healing microdreams. We invited subjects into the sleep lab to try this meditation while lying in bed and being recorded by a simple EEG headband. The instructions started with progressive muscle relaxation and establishing a feeling of being grounded (such as feeling the bed as a support), before guiding subjects into

Engineering Dreams

working with and transforming nightmares into healing imagery. The guided prompts centred on themes of feeling free from fear, feeling safe and expressing curiosity and compassion towards typical nightmare images.

The content of the resulting microdreams varied widely and arose rather spontaneously in subjects' minds during the intended sleep onset state. One subject had suffered a recurring nightmare for years of being trapped in the corner of a dark alley as a threatening man approached. She recalled that this dream began when she was young and reappeared through different periods of life, becoming more frequent and distressing at times and disappearing at others. During one session, she saw the alley scene but from a third-person perspective, witnessing it from the outside instead of being trapped in the corner, and she suddenly noticed a red fire escape next to her that she had never seen before. In a subsequent session, she decided to ascend the stairs, intending to get away, but after taking just two steps up she looked down on the man, unafraid, and he simply turned and walked away. Exploring the image through these sessions, she first found a resource of help within the dream – the fire escape – and later accomplished a sense of mastery as she literally rose above the threat. Even small changes in nightmare content like this – a new element appearing like the fire escape, or a shift in perspective from taking two steps up – can be enough to alleviate nightmares, to overwrite an old script.

Another subject, this time from the control group (composed of people who did not suffer from nightmares) recalled that as a child she had a recurring bad dream of being chased through her house and bedroom by a vicious tiger. Though she no longer had this nightmare, she explored the image in the sleep onset state and, following the prompts to reimagine the dream while feeling safe and curious, the tiger suddenly changed into a cartoon. Though she didn't know what to make of the experience initially, when she returned a week later, she stated that she had started going to bed earlier, and even her roommates noticed she was spending more time in her bedroom. She realised that ever since childhood, she had harboured some fear and discomfort around being alone in her room and had avoided going to sleep because of it, often staying up late at night and delaying going to bed until the last moment. This known phenomenon, called 'fear of sleep', often occurs in tandem with nightmares but more generally involves behavioural patterns to avoid sleep and feeling afraid of the dark or afraid of the loss of vigilance that occurs on falling asleep. In this case, working with a childhood nightmare that seemed irrelevant resolved this unhealthy sleep pattern, long after the nightmare had ceased to be a problem.

These cases, and the many that exist alongside them, illustrate the simple yet powerful potential of directing sleep onset dreams with guided verbal prompts, and reveal how these modest microdreams can support nightmare therapies and give rise to pleasant and impactful imagery.

Putting together everything we've learned so far, it's time to explore how dream engineering can enhance traditional rescripting therapy, carrying positive imagery from waking rehearsal into nighttime dreams.

Engineering Dreams

To start, recall that a typical practice of imagery rehearsal therapy asks patients to visualise positive rescripted dreams for ten to twenty minutes a day. This mental rehearsal can easily be paired with sensory cues to create a form of targeted reactivation. This was the purpose of one recent study, where patients undergoing nightmare therapy listened to auditory cues during their imagery rehearsal period; and later during the night the same sound cues were presented, via mobile app, to half of these patients when they entered REM sleep (detected by an EEG headband worn during sleep). As you might now expect, all of the patients had a reduction in nightmares following therapy, even those who did not hear the sounds during sleep – they had less frequent nightmares and less fear in their dreams after the intervention. But those subjects who were exposed to the sound during sleep improved more; they had even fewer nightmares along with more joyful dreams over time. In other words, while imagery rehearsal therapy was by itself effective in treating nightmares, the addition of targeted reactivation improved dream quality more overall, and these effects were maintained three months later.[8]

If we think back to how nightmare treatments work, in general the waking practices of exposure and rescripting help dreamers to feel less distressed by nightmares and to break free from repetitive nightmare content over time. In this study, imagery rehearsal therapy clearly helped to inhibit and extinguish fear in nightmares, reducing their occurrence. But it seems like targeted reactivation takes this a step farther – that in addition to better suppressing nightmares, the sound cues introduced more joyful themes into dreams, perhaps reactivating positive themes from patients' waking imagery rehearsal and generalising them into sleep.

To this end, other studies have also used targeted reactivation to induce more pleasant dream themes, such as, for example, flying dreams, one of the most euphoric and universal dream themes. In one study in Montreal, subjects played a VR (virtual reality) flying game in the lab before taking a nap coupled with targeted reactivation. This led to an eightfold increase in flying dreams that night, and those subjects who were exposed to auditory cues also reported more flying dreams at home over subsequent nights.[9] The researchers conducting the study also observed an unexpected correlation with dream lucidity. One subject reported their first ever lucid dream: '*Oh my God, my first lucid dream! ... I tried to make myself float a little, then once I realised that it worked, that I had control, I put my hands at my sides, just like Iron Man, and I imagined myself flying really fast ... it was incredible ... I screamed with joy as loud as I could.*' Though the VR task was not designed to induce lucidity, the experience of controlling an avatar within a virtual world could itself be a primer for lucid dreaming (indeed, video gamers have more lucid dreams than non-gamers). And flying dreams have long been closely linked with lucidity: flying is a common dream sign that clues dreamers into the fact that they are dreaming, and flying is the most frequent activity in lucid dreams. Given how euphoric and often lucid these dreams tend to be, gifting flying dreams to those with nightmares could be a worthwhile endeavour.

Beyond these attempts to induce pleasurable dreams,

targeted reactivation can also be used to support emotion regulation, the overnight therapy of REM sleep and dreams. I'll refer again to the karaoke study we encountered earlier, where scientists found that REM sleep helped to soothe the distress subjects felt when listening to their own out-of-tune singing. In this study, some of the subjects underwent a targeted reactivation protocol. The experimenters presented a scent initially when subjects listened to their karaoke tunes while awake, and later an olfactometer (a machine that triggers the release of odours) presented the same scent to subjects during sleep. Those subjects who were exposed to the scent during REM sleep were better off the next day – they had less activation in the amygdala as measured by fMRI. It is likely that smelling the familiar scent during REM sleep reactivated their memories of out-of-tune signing, and thanks to REM sleep's emotion regulation function, this dampened their negative reaction to the embarrassing tunes the next day.[10]

Taken together, targeted reactivation can bolster rescripting therapy, instill more pleasant and lucid dreams, and enhance the benefits of REM sleep for dampening distress – the overnight therapy of sleep.

From another angle, targeted reactivation has also been used to selectively *weaken* memories. This is referred to as 'targeted forgetting', where pieces of memory are abolished during sleep. As you can imagine, this could prove very useful in treating post-traumatic nightmares, where there is a need to extinguish or to forget aspects of a traumatic memory, particularly its negative tone (though this kind of research is still in early stages). To give an example of targeted forgetting: when subjects were instructed to intentionally memorise or forget certain parts of visual scenes, pairing sound cues with the 'forget' instruction and replaying these sounds during sleep resulted in even worse memory for the visual scenes after sleep. Impressively, targeted forgetting can even change post-sleep behaviours, as seen in one study that sought to eliminate a particular undesired behaviour: cigarette smoking. When sixty-six smokers spent the night in a sleep lab in Israel, unbeknownst to them, the odour of cigarettes was presented alongside profoundly unpleasant scents, like that of rotten eggs. This created conditional learning where the smell of cigarettes, and the behaviour of smoking by extension, became associated with rotten eggs. Subjects exposed to these scents during sleep (but not during wakefulness) smoked less in the days following the procedure, compared to the days before.[11]

While all of this research is at this point still rather preliminary, targeted reactivation (and targeting forgetting) shows promise as a method to enhance nightmare therapies, be it to induce more pleasant dreams, dampen post-sleep distress, or even abolish traumatic memories or unhealthy threat responses. All in all, these varied dream engineering approaches of targeted reactivation and targeted incubation are proving useful for nightmare treatment, providing patients with more lucid and aesthetic dreams and microdreams and giving dreamers more control over their dream worlds. These methods seem well poised to supplement and enhance traditional approaches to treating nightmares.

Interfacing with the dreamer

While the engineering techniques discussed so far rely on a mix of presleep training paired with within-sleep stimulation, there are also ways of directing dreams using within-sleep stimulation alone. These tend to have less precise or pronounced effects on sleep and dreams but in a way are less labour-intensive, steering dreams without any need for prior practice or waking effort.

Let's take a look at how some different types of sensations can shape dreams implicitly within sleep – and whether these could also be useful in nightmare therapy.

In some ways, scent is an ideal sensory stimulus for dream engineering because it is one of the only sensations that does not typically disrupt sleep. Whereas sounds or lights can easily startle you into wakefulness, odours are less arousing. Unlike most sensory information, odour perception does not require any processing by the thalamus, a part of the brain linked to arousal; rather, olfactory information passes more directly to the sensory cortex where it can be perceived without thalamic input, meaning it is less arousing than other senses. The olfactory system is also closely linked to the brain's emotion processing areas, such as the amygdala and hippocampus, which are integral to forming emotional memories. For this reason odour is particularly relevant to, and useful for, nightmare treatments.

Already, the use of pleasant scents has been explored as a means of improving both sleep quality and dream intensity in patients with PTSD. In one study, after patients selected their favourite scent from orange, peach, rose, or lavender, presenting this scent during five nights of sleep at home (delivered by a wearable device) resulted in a lower dream intensity, compared to patients receiving a placebo. In this case, the scent is presumably acting in a manner similar to aromatherapy, calming dream intensity via the calming effects of the odour. Similar research has shown that lavender improves subjective sleep quality, and the scent of roses boosts emotionally positive dreams (whereas the smell of rotten eggs led to more negative dreams). Smells can also be used to entrain rhythms of the sleeping brain; odours such as vanilla or lavender have been shown to amplify slow waves when presented during deep sleep, which is often a sign of improved sleep quality. In general, while these effects are not immense, introducing pleasant scents into a bedtime routine is a relatively simple way to quiet the dreaming mind and even improve sleep quality in patients with PTSD, supporting a good night's rest and recovery.

A related technology designed to alleviate nightmares relies on vibration rather than odour. An app called Night-Ware, approved by the Food and Drug Administration in 2020, claims to detect and interrupt nightmares when they occur. This app relies on a watch that measures heart rate, body movements and body position to calculate a 'stress index' during sleep, and when the app suspects the user is in the midst of a nightmare, it presents a subtle vibration at short intervals to disrupt this state. Thereafter, when the stress index returns to lower levels (either within sleep or due to an awakening), the vibration ceases. In a clinical trial with seventy-two PTSD patients, those who used the watch reliably – at least fifteen nights in a thirty-day period – reported improved sleep and nightmare reduction overall, compared to those who did not. Though there were no objective measures of sleep, it's thought that perceived sleep quality improved due to having less distressing awakenings from nightmares, even if the vibrations provoked minor arousals.[12]

While scent subtly soothes sleep and dreams without being arousing, vibration is more abrupt, interrupting nightmares in part precisely because it *is* arousing.

Another sensation to consider in this context is temperature: sleep is closely tied to fluctuations in core body temperature over the course of the twenty-four-hour day, and both sleep onset and sleep depth are sensitive to environmental changes in temperature. Because of this, warming sleep masks and heating and cooling sheets and mattresses have been designed to promote better sleep. For instance, masks that warm the eyes just prior to sleep led to improved sleep quality and even enhanced slow waves in deep sleep. More generally, warmth seems helpful for falling asleep, whereas a cooler body temperature is preferable for maintaining deep sleep. Smells can also produce sensations of heating or cooling, specifically those that activate the trigeminal nerve, such as capsaicin or peppermint. But despite the age-old anecdotes of fever dreams (Could nightmares be produced by heating in the brain?), relatively little is known about how temperature affects dreams. Still, approaches to modulate body temperature and regulate sleep across the night could prove worthwhile for patients with nightmares and PTSD.

In the realm of lucid dreaming, so far simple visual and auditory cues seem most reliable, or at least the most studied, for finding their way into dreams and helping dreamers become lucid. But what other types of stimuli could be used to induce lucidity? Scientists at the Institute of Sport Science in Switzerland are looking into electrical muscle stimulation (small electric shocks that cause muscle twitches) or vestibular stimulation (an electric current applied behind the ears that creates the feeling of falling or spinning) as methods to trigger lucid dreams. Preliminary results so far suggest that muscle stimulation is effective with 50 per cent of subjects reporting successful lucid dreams (although only 25 per cent performed clear eye signals); whereas vestibular stimulation did not seem to work, possibly because this sensation is too arousing or disruptive to sleep.[13] Odour, on the other hand, may not be arousing enough: in one study when an odour was paired with reality testing while awake, then re-presented during REM sleep, not one of the sixteen subjects had fully lucid dreams, meaning none of them were able to signal that they were lucid with left-right eye movements. So there's clearly a balance to be found between using sensations that are arousing enough to get into dreams and provoke lucidity but are not so intense as to disrupt sleep altogether.

In a more extreme form of dream engineering, some researchers have looked at the possibility of using electrical brain stimulation to increase dream lucidity. As we know, the brain cycles through different rhythms of electrical activity during sleep, and in lucid dreams, we think that these rhythms become faster and more wake-like, especially in the frontal cortex. Scientists can apply electrical stimulation to the frontal cortex to increase wake-like brain activity in these regions (similar to lucid dreaming) and in theory induce more lucid dreams. Results so far are mixed, though it is a fascinating area of study. Other related work has applied methods of entrainment to enhance slow waves in NREM sleep (via electrical stimulation or pulsating auditory cues, for example) and this can actually suppress dream recall, though whether it could similarly reduce nightmares is not yet known.

In sum, various types of stimulation exist for engineering dreams, with diverse effects on underlying dream content. Subtle scents and temperature can foster more mild or pleasant dreams, or restore sleep quality in patients with nightmares. Audiovisual cues and vibration are generally more arousing, which is a plus when it comes to inducing lucid dreams and interrupting nightmares. And while electrical brain stimulation can reliably entrain brain waves during sleep, the effects on dreaming are so far less consistent and not well known. In general, these approaches can offer a complement to traditional nightmare therapy, and subtle effects can be seen from something as simple as adjusting the aroma or temperature of the bedroom.

So far we've seen a number of methods to push and prod at the sleeping body, in the quest to influencing dreams. Now it's time to look the other way, to see how dream engineering can be used to better *record* dreams. Remember that the bodily interface works two ways: engineers can stimulate the sleeping body, but dreamers can also control their physical bodies from the inside, sending information in the other direction, to the outside world, while remaining asleep and adream.

So how can we listen to these seemingly silent sleeping bodies, reading out physical memos from the metaphysical dream world?

In what is perhaps one of the most incredible recent discoveries relating to dream science, scientists have now shown that they can carry on two-way dialogues with lucid dreamers, conversing across the barrier of sleep. An international study, led by teams in the United States, the Netherlands, Germany and France, tasked lucid dreamers with listening for speech in their dreams and answering questions while asleep.[14] In some cases, experimenters presented softly spoken math problems over bedroom speakers to the lucid dreamers, who were instructed to respond with the correct number of left-right eye movements. One lucid dreamer reported hearing the math problem 'What's 1 plus 2?' in his dream through a car radio and responded accordingly with three left-right eye movements, which were confirmed by the experimenters. Other lucid dreamers were asked to respond to yes-or-no questions with one of two body signals – either frowning their eyebrows or smiling. One subject dreamed: '*I was at a party and ... I heard your voice as if you were a god ... coming from the outside, just like a narrator of a movie. I heard you asking whether I like chocolate, whether I was studying biology, and whether I speak Spanish. I wasn't sure how to answer the last one, because I am not fluent in Spanish ... In the end, I decided to answer 'NO' and went back to the party.*' The frowning signal confirmed their dream response, visible to experimenters in the real world.

Overall, in almost half of the lucid dreams, subjects were able to complete this dialogue at least once; but this accounted for only 18 percent of the total communication attempts made by the experimenters, suggesting that real-time dialogue, while viable, requires further fine-tuning.

The fact that lucid dreamers can comprehend speech from within sleep is remarkable, as this requires quite a high level of cognitive function. Importantly, it means that we don't need to rely on complicated codes like flashing lights or beeping sounds to communicate with dreamers. Speech simplifies the process of getting messages into the dream world, especially compared to earlier work where dreamers went so far as to learn Morse code, in order to interpret a series of beeps while in their dream.

But what other ways are there for getting messages out of the dream world? Is there an easier way to communicate than left-right eye signals or yes-or-no muscle twitches?

This is the focus of recent work by the team at the Institute of Sleep and Dream Technologies in Germany. If you remember, I described earlier that lucid dreamers can trace shapes with their eyes, by holding their thumb out in front of them and following its movement. Well, this method could be used to send eye-movement messages from a dream; for instance, one could trace the shapes of cursive letters drawn with their thumb, writing in the dream sky – 'Hello, dream!' – for experimenters to read. While so far only proof-of-concept examples exist, the hope is that with further study, we will improve methods for dreamers to relay more complex messages that can be decoded from their dreams.[15]

While it is still early days, my hope is that communication like this could eventually be helpful for nightmare sufferers. This was the case for one subject who took part in a lucid dreaming study at Northwestern University. The subject had weekly nightmares at home, and unfortunately also had a nightmare while in the lab. When she realised she was dreaming, she reached out to the experimenter with a left-right eye signal because she knew they would wake her up for a dream report. She essentially used the eye signal to call for help and later reported, 'In the dream I was aware that you (the experimenter) were there and reachable.' This experience of being able to communicate and be pulled out of her nightmare had a significant impact – in this case, her nightmare frequency decreased in the month following the study. Since feeling helpless is such an exacerbating feature of nightmares, having a metaphorical helping hand, a tangible means of getting out of a nightmare, can be powerful. Imagine if nightmare sufferers could trigger an alarm clock or call a friend to help them wake up, or even converse with a therapist while the nightmare is ongoing. These possibilities are for now still in the realm of science fiction, but conceptually they are all feasible.

Finally, it's worth noting that clinicians are increasingly turning to other modern technologies to supplement dream therapies, especially VR. In a way, VR is similar to lucid dreaming in that you are immersed in a virtual world and yet are aware that what you are experiencing is 'not real', so to speak. Because of this, VR has been used to train subjects to lucid dream, by having them practise reality testing in games designed to be dreamlike – with characters suddenly transforming or gravity failing, for example. In these VR games, subjects can more genuinely question whether what they are experiencing is like a dream, whereas the waking world is not often dreamlike. When subjects practised reality testing in VR a few times a week for four weeks, this more than doubled their incidence of fully lucid dreams compared to subjects who practised only in the real world. VR has also been used to enhance rescripting therapy, where in one study patients practised transforming nightmarish images by drawing over them or changing their size or shape. After two sessions per week for four weeks, these patients had decreased nightmare distress and lower waking anxiety and PTSD symptoms.

Just like dreaming, VR provides a multisensory and immersive experience, one that engineers can to some extent design, in the quest to help dreamers realise more lucid, aesthetic and pleasant dreams.

Beyond the lab

This is all very well, but most people don't have access to a laboratory, so a key area of work for translating these benefits to the public is the development of dream engineering devices that can be used at home. This was one of the primary motivators for the initial launch of the dream engineering research network, a group of scientists and engineers devoted to better influencing and recording dreams within and beyond the laboratory. In 2018, as a dream scientist myself, I was working almost exclusively on laboratory-based PSG (polysomnography) studies when I discovered a group of engineers who were creating exciting new devices designed to measure sleep and dreams at home. The Fluid Interfaces Group, led by Pattie Maes, was building wearable sleep masks, glove-like sleep-onset detectors, and clip-on olfactometers, all designed to interface with dreams in the real world. While working with this team, I led the first international workshop on dream engineering at the MIT Media Lab in 2019, bringing together over fifty leading dream scientists and developers in the field.

Several international collaborations have grown out of this network, including multisite studies designed to test and fine-tune dream engineering tools in larger samples of subjects. And collaborations with industry aim to validate new sleep apps and wearable devices, to see whether we can better record dreams at home. It is increasingly hard to keep apace of all the developments in the field, as seemingly every day new sleep trackers hit the market, including mobile apps and wearable devices like the Oura Ring or Muse headband. While all of these offer wireless sleep recording at home, the challenge of real-time sleep staging – algorithms to detect sleep stages without delay – makes these devices less reliable than laboratory PSG. Nevertheless, wearables can provide some index of sleep quality at home, and sleep staging methods are improving rapidly.

To give a few examples of devices my team and I have tested in our studies: the ZMax is one wearable headband that has been designed with dream engineering in mind. It's a simple EEG headband that can detect REM sleep and deliver sensory cues in real time (with LED lights and vibration built into the headband). The device is quite useful for experiments on lucid dream induction, though it requires some technical skills to use. Other commercial devices, such as the Muse or Dreem headband, come with more user-friendly apps to track your sleep metrics at home, but without the option for automatic sensory stimulation (for now). Mobile apps often rely on data from a Fitbit or Apple watch to identify sleep stages in real time, or use data from the mobile phone itself to detect sleep. One such app used the phone's accelerometer to detect body movements as a proxy for wakefulness and sleep; the app first delivered presleep lucid dream training paired with auditory cues, and re-presenting these audio cues during early morning sleep led to increased rates of lucid dreaming.

Besides inducing lucid dreams, devices are being designed to record features of dream content, especially by detecting levels of arousal or emotion in the body during REM sleep. For instance, scientists in Cyprus are creating machine-learning algorithms to detect nightmares, based on EEG measures from a wearable headband. As mentioned earlier, the FDA-approved prescription app, NightWare, uses biosignals to detect physiological stress during sleep and triggers a light vibration to awaken patients or interrupt a nightmare before it peaks. Other research has uncovered EEG signatures that seem to precede the onset of lucidity, specifically an increase in wake-like brain activation just before a lucid dream. If we know what the brain looks like just prior to lucid dreaming, we can pinpoint the best time to deliver lucid induction protocols – such as presenting sensory cues when the brain is in a prime state to perceive these cues and become lucid. Altogether, these approaches aim to identify opportune moments to intervene in dreams, based on bodily signals and sleep stages, to interrupt nightmares or induce lucidity all within the comfort of home.

While it is still early days in the development of such dream aids, it is clearly a fast-growing market, and sleep medicine is quickly integrating such tools into clinical practice, including into the treatment of nightmares. But could sleep trackers have a downside?

Indeed, there is some call for caution in the increasing use of sleep technologies. Dream scientist and philosopher Elizaveta Solomonova in Montreal has discussed the potential nocebo effect of using sleep health wearables. The inverse of the 'placebo' effect, the nocebo is essentially an unintended adverse outcome that results from using a medicine or device, which stems from a person's negative expectations about how the medicine or device will work (whereas with the placebo effect, simply expecting a medicine to work leads to better outcomes).

In the case of sleep tech, using wearable devices can actually worsen one's sleep perception and have other consequences on waking health. In a way, sleep becomes viewed as an achievement, and if an app tells a user they missed their deep sleep or did not get a full eight hours, that person is likely to feel dissatisfied or even more tired in the morning. In experiments where researchers manipulated wearable devices to deceptively display poor sleep scores, subjects had worse cognitive performance and poorer mood the next day (even when their sleep quality was good), compared to subjects receiving better sleep scores. And unfortunately, sleep trackers are often inaccurate, especially for those with unhabitual sleep patterns, since algorithms are mostly trained on people with healthy, 'normal' sleep. In this context, overuse of sleep trackers could cause more harm than good.

Engineering Dreams

Besides the concerns around sleep tech, scientists have raised similar questions over the potential side effects of controlling dreams, and especially of lucid dreaming. Many of the benefits of lucid dreaming seem to depend on two key factors: first, being able to successfully induce lucid dreams, and second, being able to successfully control lucid dreams. Said another way, trying and failing to induce lucid dreams, or having lucid dreams with very little control, can have unintended adverse effects.[16]

The most likely negative consequence of practising lucid dreaming techniques is sleep loss. Indeed, people who frequently attempt to induce lucid dreams have worse sleep quality, although lucid dreaming itself does not seem to be the culprit in this link. Consider the wake-back-to-bed method, where sleep is intentionally disrupted in the early morning. If someone repeatedly tries this technique and fails to fall back to sleep, this will clearly result in sleep loss over time. This is not so much a problem when someone succeeds in having a lucid dream, meaning they have successfully returned to sleep. In fact, when subjects do have lucid dreams, sleep quality does not seem to be adversely affected, and there is some evidence that nights with lucid dreams are associated with better sleep quality overall. Nevertheless, there is a subset of highly frequent lucid dreamers who describe their lucid dreams as tiring, feeling unable to rest during sleep because of 'vivid lucid dreams that are so intense I feel exhausted when I wake up'. This seems to occur only rarely in expert lucid dreamers and is unlikely to develop through novice lucid dream training. But for those who suffer from repetitive and tiring lucid dreams, a form of dream therapy for calming dream intensity, and even rescripting dream content, could be warranted.

In general, these cases raise some cautions around lucid dreaming therapy: that it may not be appropriate for all patients and that induction techniques, like wake-back-to-bed, may be preferable in small doses. In fact, wake-back-to-bed is also reported to result in more frequent false awakenings, which are experiences where one believes they have woken up, only to later realise they were still dreaming. Some dreamers report having multiple false awakenings in a row, repeatedly thinking they have woken up, only to later awaken for real. This can cause distress and dream-reality confusion – the dreamer uncertain whether or not they are fully awake. Nevertheless, false awakenings can also be used as a reliable dream sign – and therefore aid in lucid dreaming. Experienced lucid dreamers will often complete a reality test upon waking, especially if it's the morning after a lucid dream induction attempt. They might try to turn on a light or breathe through a plugged nose – and if the test fails they know they are still dreaming. In this way, false awakenings can serve as a reliable bridge to lucidity rather than a negative outcome of failed induction.

Somewhat related to false awakenings, sleep paralysis can also occur after wake-back-to-bed periods, which is not surprising because of the quick transition into REM sleep that occurs in the morning. Sleep paralysis can be somewhat similar to lucid dreaming in that the dreamer is often aware that their body is still asleep, but the inability to fully wake up makes it a frightening experience. As with false awakenings, experienced lucid dreamers can learn to use sleep paralysis as a gateway into lucid dreaming. In order to do so, rather than trying to wake up, the dreamer can relax and focus on calming their breathing, falling deeper into sleep while maintaining awareness. Viewed this way, learning to lucid dream can actually help those with sleep paralysis and false awakenings, in a manner similar to nightmare therapy – transforming these disturbing experiences into lucid and positive dreams.

Now, once lucid, there is still the possibility that a dream will present threatening or fearful imagery, sometimes beyond the dreamer's control. This can develop into what's called a 'lucid nightmare', where a dreamer is aware that they are dreaming but unable to control a frightening dream. The dreamer typically loses some lucidity as they become convinced that the threat is real, and as emotions intensify, it becomes increasingly difficult to remain lucid or to control the dream. Thus, a first step to counter the development of lucid nightmares is to practise maintaining or returning to a calm and relaxed state of mind, before tapping into some of the lucid control strategies described previously, such as using one's attention, voice and actions to help tame an unpleasant dream. Training in these techniques of staying calm and exerting control over oneself and the dream environment are essential to any sort of nightmare therapy, to prepare patients to manage negative dream contents that can arise once lucid.

A final concern voiced by clinicians is that lucid dreaming may not be advised in certain patient populations, particularly those prone to psychosis.[17] This is because induction techniques like wake-back-to-bed and reality testing are to some extent designed to blur the boundaries between sleep and wake states – interrupting sleep with wakefulness and questioning whether one is awake or dreaming throughout the day and night. This deliberate confusion of reality and dreaming is ill-advised for those at risk of psychosis. In fact, frequent use of lucid dream induction techniques has been associated with an increase in dissociative symptoms – such as distancing oneself from reality to avoid emotional pain. Given that a subset of patients with PTSD tend to experience dissociative symptoms (about 15 per cent), lucid dream induction techniques like wake-back-to-bed or reality testing may not be recommended for these particular patients. Nevertheless, other techniques, such as targeted reactivation or simple intention setting, could allow patients who may be susceptible to dissociative symptoms to still experience lucid dreaming without the risks associated with repeatedly disrupting sleep-wake states.

Overall, it is important to further discern when and for whom lucid dreaming is an appropriate therapy, and to develop techniques for more reliably inducing lucid dreams without sleep loss, as well as skills for controlling dream content, which will further the benefits to be had from this kind of therapy. In particular, instructions in how to manage dream emotions and constructively engage with dream content can help to calm the mind in overactive lucid dreams, relax in the face of sleep paralysis episodes, or even transform dreams when lucid dysphoria creeps in. These experiences will contribute to a greater sense of agency within dreaming, a contrast to feeling helpless or at the mercy of disturbing dreams and nightmares.

*

Engineering Dreams

In the end, dream engineering offers a fascinating new avenue to designing our nightly dreams. Building on the science we saw earlier, where we learned that dreams are real embodied experiences, engineers are now learning how to shape and mould these multisensory dream worlds on demand. Simple things like scent cues can soothe dream mood; rhythmic stimulation can slow brain waves or calm dream intensity; and all forms of targeted reactivation can bridge waking interventions into sleep – for instance, sending sensory cues into the dream world as reminders that one is dreaming. These tools intersect with the natural design of the dream world. And their chief and most interesting use, in my work at least, is for healing nightmares and promoting more lucid and positive dreaming.

Altogether, dream science has come on leaps and bounds in recent decades, and with novel, cutting-edge technologies, which are developing rapidly, there is a whole new world of possibilities for dream intervention, to combat bad dreams and improve sleep and well-being. This becomes even more important when we see just how prevalent nightmares are in all manner of sleep and psychiatric disorders, including PTSD, and also insomnia, psychosis, addiction and chronic pain, to name a few. Given how disruptive nightmares can be to health, improving access to treatment, perhaps with some of the novel tools outlined here, is essential, even if there are still open questions about what works best and why. This is especially critical given how rare it is for nightmares to be treated: less than a third of patients report their nightmares to clinicians, and most clinicians do not ask about nightmares or know how to treat them.

Into the Dream Lab

Already we know that treating nightmares can improve the lives of people who suffer from them, and we'll see next how these therapies are proving ever more valuable for the many patients who have nightmares alongside other sleep and mental health disorders, too.

Part IV

Where Else Is Dreaming Relevant?

7

Bad Dreams and Health

In the summer of 2012, I attended the International Association for the Study of Dreams annual conference for the first time, set across the San Francisco Bay in Berkeley, California. I was slated to give a scientific presentation, the first of my graduate career, on a study of bad dreams and nightmares. I remember feeling especially nervous to be speaking to this crowd, full of scientists and writers whose work I followed and admired. I over-prepared and memorised everything I planned to say, even scripting answers to the questions I thought I might get asked.

I'll never forget the first question I was asked: 'Why do you call them *bad* dreams?,' one woman critiqued. 'They're not *bad*!' Well, this was not a question I had prepared for, and it's one I've contemplated many times since.

In the most basic sense of the term, 'bad dreams' are simply negative dreams that do not provoke an awakening because they are less intense than nightmares. But are bad dreams really bad? If so, when and for whom? And can bad dreams be good for us?

While much of my research is designed to dissect just how and when nightmares are bad for us, it is also clear that milder forms of bad dreams are a common and normal response to the stresses of waking life. The content of bad dreams varies from merely negative to macabre, bizarre, or disturbing, and what's especially interesting is how particular bad dream themes appear alongside changes in our waking health: drug-using dreams in addiction, aggressive fight-or-flight dreams in Parkinson's disease, and more. Physical changes such as pregnancy or sensory loss are also linked to specific dreaming patterns that remind us that our bodies play a crucial role in shaping the dreaming mind.

Today, we know that bad dreams occur throughout the lifespan; they come and they go, often spurred on by other sleep health, mental health and even physical health conditions. However, modern medicine has largely neglected dreaming. The field of sleep medicine has favoured more objective and physical measures of sleep in lieu of attention to the sleeping mind; and even psychiatric medicine has mainly concerned itself with the waking mind, with disregard for our experience in sleep. Any consideration of dreaming has been especially absent from physical medicine, where dreams have been deemed largely irrelevant.

Because of this, many colleagues and I are part of a recent push to reframe dreaming as an essential and uniquely informative component of sleep, physical health and mental health across the lifespan. We believe that how we experience sleep is not only relevant but vital to our general fitness and well-being. Indeed, accumulating evidence is now converging on the fact that dreaming is directly tied to multiple facets of health and is a useful tool that has diagnostic, prognostic and therapeutic potential across medical fields.

Dreams and sleep health

It might be surprising to learn that, aside from nightmares, attention to dreaming in clinical sleep medicine has been nearly absent since the field began in the 1960s. Sleep medicine has grown as a predominantly behavioural field of medicine, with a focus on objective measures of sleep. Even self-reported estimates of sleep duration and quality are typically less trusted by clinicians than measures obtained by polysomnography. There has been scant attention to subjective experience – how we feel during sleep and how we dream – despite clear evidence that how we think and feel in waking life is strongly tied to our health.

So why this neglect of dreaming? On the one hand, sleep medicine is a relatively young field. The first sleep clinics in the United States came about in the late 60s and 70s, though medical training in sleep wasn't required until decades later, and even today physicians obtain only a few hours of sleep training in medical school. While sleep medicine has made enormous strides in bringing sleep health into public awareness in just a few decades (almost everyone has a general concept of sleep hygiene and the 'eight-hours-a-night' guideline), there are still gaps in understanding. In my opinion, a necessary next step is to look at how patients *feel* during sleep and how they dream. In recent years, the Society of Behavioral Sleep Medicine has begun to promote the study of dream interventions and to expand sleep medicine inward, to the sleep experience.

In fact, while nightmares are relatively well-studied as a standalone sleep disorder, there are many other sleep disorders that are associated with bad dreams, especially REM parasomnias, that is, disorders affecting REM sleep. These disorders mainly involve abnormal and undesirable physical behaviours that arise during REM sleep, but increasingly we are seeing that disturbed dreaming is a key symptom worth studying and treating, too.

To begin with a powerful example: REM sleep behaviour disorder (RBD) occurs when the normal paralysis of REM sleep breaks down, causing patients to physically act out their dreams. RBD affects less than 1 per cent of the general population, but it often develops somewhat suddenly in men over sixty, with about an 8 per cent prevalence in this population. The clinical significance of RBD cannot be understated: up to 90 per cent of patients with RBD develop a neurodegenerative disease within a decade of diagnosis, most commonly Parkinson's disease.

A primary symptom of RBD is the occurrence of vivid and repetitive nightmares that are enacted during the night. These are often violent or aggressive dreams with behaviours that mimic fight-or-flight responses, such as grabbing, punching, kicking and flailing. Patients often seek help following an injury to themselves or their partners. The dreams and behaviours have a striking and sudden onset, and although patients sometimes present other mild symptoms, such as a loss of smell or mild memory impairments, RBD itself is the clearest early sign of neurodegeneration. The aggressive dreams and behaviours seem to reflect the early stages of neuronal loss – specifically the loss of brain cells in both the brainstem and a region called the substantia nigra, where movements are in part controlled. With the loss of these neurons, the paralysis of REM sleep is disrupted; and the brain loses the ability to regulate dopamine, a neurotransmitter essential for motor control and for the fight-or-flight response that kicks in when we perceive a threat. In RBD, these brain changes seem to result in excessive and exaggerated muscle twitches during REM sleep, alongside aggressive fight-or-flight dreams that erupt into repetitive kicking or punching behaviours.

While treatments for RBD have focused on pharmaceuticals designed to suppress REM sleep (which we'll come back to later), one case report explored the possibility of treating the nightmares psychologically, in order to reduce dream enactment behaviours and ease sleep. Physicians delivered five sessions of imagery rehearsal therapy over five months to a patient with RBD, which decreased the frequency and intensity of their bad dreams and reduced the reported number and intensity of dream enactments, while improving sleep quality, too. In other words, by treating bad dreams psychologically, the physical acting out of dreams lessened, which makes sense when remembering that dream enactments usually arise during emotionally intense dreams.[1]

While studies of nightmares, and nightmare therapy, in RBD are still nascent, the implications are powerful. Given that aggressive dreams are one of the earliest prodromes (pre-disease symptoms) of Parkinson's, and knowing that sleep is crucial for cognitive function, it is possible that treating and improving sleep quality early in disease progression could help to stave off cognitive decline.

Turning to another REM parasomnia where disturbed dreaming and loss of motor control are central: patients with narcolepsy often experience nightmarish imagery and undesired muscle paralysis on the threshold of sleep and wakefulness. Narcolepsy is caused by the destruction in the hypothalamus of orexin neurons, which normally stabilise the brain's switch between sleep and wake states. Patients sometimes fall asleep at inappropriate times, like at the dinner table or while driving, and suffer extreme sleepiness during the day. They can also experience something called 'cataplexy' – the sudden loss of muscle tone while awake, which can be triggered by strong emotions. The typical diagnostic test for narcolepsy is to invite patients into the clinic and observe how quickly they fall asleep during multiple short nap periods scheduled across the day. They will often enter REM sleep quickly, within five minutes of falling asleep, and report frequent sleep paralysis episodes, with muscle paralysis and nightmares emerging during transitions between REM sleep and wakefulness. As we saw earlier, these episodes often feature the sense of a threatening presence in the room, a pressure on the chest and even feelings of suffocation.

Besides sleep paralysis, over a third of patients with narcolepsy suffer from nighttime nightmares – six times higher than in the general population. And patients also have generally more negative and aggressive dreams, including dreams with aggressive sexual themes. It has long been known that disturbing dreams cause distress for those with narcolepsy, but up until recently, it was unknown whether these kinds of nightmares could be responsive to traditional therapies. In a clinical trial at Northwestern University in 2024, six patients underwent a combined imagery rescripting and lucid dreaming therapy for nightmares. During one session, patients actually came into the sleep laboratory and underwent a targeted reactivation procedure, where sound cues were linked to imagery rehearsal and lucid dream training while awake, and the sounds were then replayed during sleep. At the end of therapy, in only a matter of weeks, all of the patients had less severe and less frequent nightmares. This was quite transformative for these patients who had suffered recurring nightmares for decades, and yet had never been asked about or offered treatment for their nightmares. As one patient reported, 'I didn't think it was going to work, but it was amazing. I am so blown away how well rescripting works.' The results are highly promising and highlight the importance of studying and treating nightmares across sleep disorders.[2]

In addition to nightmares, individuals with narcolepsy also experience something called 'dream-reality confusion', where they are unsure of whether a memory occurred in waking life or in a dream. Occasional dream-reality confusion is quite normal, but those with narcolepsy experience it more often than is usual, likely because their dreams are so vivid and realistic and frequently intrude into wakefulness. For example, case reports have described false accusations of sexual assault stemming from dream-reality confusion, when a patient mistook a dreamed assault for an actual event. While there are no existing treatments for dream-reality confusion, one possibility is that simply increasing awareness of one's dream patterns, through dream journaling or lucid dreaming, could improve memory for when events occurred in dreaming or in waking life.

The prevalence of nightmares featuring sexual aggression in narcolepsy is perhaps surprising, but it is likely due to the natural physiology of REM sleep. We know that physiological sexual arousal is present in over 90 per cent of REM periods, and although dreams are only sexual about 15 per cent of the time, this percentage is much higher (up to 80 per cent) when subjects are awakened specifically during periods of increasing sexual arousal in sleep. Lucid dreaming research has even shown that dreams ending in orgasm correspond to real physiological orgasm in the body. In general, this is just another example of how closely tied REM sleep is to our physical functioning, stimulating fight-or-flight dreams, erotic dreams and an array of other embodied and survival-oriented themes. These bodily-based dreams can become aggravated in different sleep disorders and even enacted through behaviours, as we'll continue to see next.

Turning to the umbrella of NREM parasomnias, these are sleep disorders where people physically act out different behaviours usually during deep sleep. The most common examples are sleepwalking and sleep talking, but others include sleep-related eating, sexsomnia, sleep terrors and mild confusional arousals, also known as 'sleep drunkenness'.

As their names suggest, patients display these wake-like behaviours during sleep, ranging from talking, walking, eating, sexual activities, or even more complex behaviours like playing a musical instrument. More extreme behaviours can result in injury, from accidents like running into walls or furniture, jumping out of windows, recklessly driving an automobile, wandering around streets, walking into lakes, climbing ladders and so on. Despite being asleep, patients can display substantial dexterity, which has been noted in numerous legal cases of sleep-related violence, where patients have been known to wield and even aim weapons such as loaded shotguns.

The most common and relatively safe behaviour is sleep talking, which is in fact so common that it is rarely considered a disorder unless it severely interrupts sleep. The majority of sleep speech is pretty simplistic, mostly expressions of negation (e.g. 'no,' 'uh-uh,' 'stop!'); the brevity of speech is likely in part due to muscle inhibition during sleep, meaning mouth movements are limited. In cases of more elaborate sleep talking, the speech seems to remain grammatically correct and the sleeper sometimes even respects turn-taking, leaving a gap for their imagined partner to answer. In general, the few studies of sleep talking suggest the content of sleep talking can be related to recollected dream content, and in some cases sleep talking even reflects material learned prior to sleep, presumably revealing how memories are being processed in the sleeping mind. New approaches are attempting to decode more subtle facial electrical signals that are present in speech muscles, even in the absence of outward sleep talking; this could pick up on more subtle forms of dream speech as a future avenue for listening in on the dreaming mind.

While it might seem intuitive that sleepwalking, too, would be related to dream content, up until very recently clinicians actually thought that dreaming was absent during sleepwalking episodes. In fact, the diagnostic criteria for sleepwalking requires that patients have limited or no recall for their experience after an episode. When these diagnostic criteria were first established in the 60s and 70s, most sleep scientists thought dreaming was not possible in NREM sleep: because sleepwalking occurs during NREM sleep, patients must not have been dreaming. In the decades since, not only has dreaming been largely ignored in the study of sleepwalking, but patients who *did* report dreaming were misdiagnosed, classified as not having a NREM parasomnia because they recalled dreams during episodes.

I remember that early in my graduate career, I attended a talk by dream scientist Antonio Zadra, now a colleague of mine in Montreal; his work showed unequivocally the presence of dreaming in sleepwalking patients, and his research was some of the first to convince clinicians that dreaming is relevant to our understanding of these disorders. Now, we know that up to 80 per cent of patients recall some dreaming after episodes, although reports can be quite brief, and patients do experience some confusion and amnesia on awakening. Patients often qualify these brief dream experiences as nightmares, short unpleasant scenes of threat or danger.

Research using high-density EEG shows that these patients are in a sort of hybrid state, half awake and half deeply asleep in different parts of the brain. Brain recordings of this 'hybrid state' align with patient reports that they sometimes see their dreams overlayed on real physical settings, and their overt behaviours coincide with this mixed dream-reality perception (a patient might report seeing bugs crawling in bed, after frantically brushing at the covers). Though recordings of such behaviours in the laboratory are still scarce, reported behaviours seem to be repetitive within patients (for instance, a sleep-eater will seek and consume food each episode but not switch to other behaviours). While it's not well understood what causes these behaviours, there is evidence that some cases are linked to psychological issues in waking life. For instance, sleep-related eating has been linked to disordered eating in childhood. If sleep-related behaviours *are* tied to underlying psychological issues, or even to repetitive dream scripts, then dream interventions could help, resolving bad dreams and the corresponding behaviours, too. While dream-specific interventions have not yet been studied, Zadra's recent work suggests that hypnotherapy is effective – this involves a few sessions of relaxation and positive imagery rehearsal of sleeping peacefully, which is then practised nightly before bed, in a manner similar to other imagery-based therapies.

A more extreme form of NREM parasomnia occurs in the case of sleep terrors, which occur during confused partial awakenings from NREM sleep. During sleep terrors, patients express extreme distress through facial expressions, screaming, or agitated movements like hitting the wall or flailing as if in imminent danger. These actions are more vigorous and frantic than typical sleepwalking. Like sleepwalking, sleep terrors are more common in children and become less prevalent with age. Up to 17 per cent of children sleepwalk (at age eleven or twelve), and while most outgrow this behaviour, it persists in adulthood in up to 25 per cent of cases (1–4 per cent of adults overall). Sleep terrors occur in 18 per cent of five- to seven-year-olds, whereas only 2 per cent of adults report sleep terrors. A child in the midst of a sleep terror might cry intensely and display overwhelming terror or even pain. The fact that they are inconsolable in this state can be a huge source of distress for parents hoping to comfort them.

While dream content is difficult to collect from children after sleep terrors, studies of adults show that dreams can and do occur in this state and can range from brief frightening images or thoughts to more elaborate and narrative dreams. It may seem intuitive that some mental content coincides with this clear state of terror, but again, because they occur during NREM sleep, historically scientists have assumed that dreaming was not possible during sleep terrors. Instead, it was thought that sleep terrors were a purely physiological phenomenon, and still today, research on dreaming in sleep terrors is scarce and psychological treatments for sleep terrors are practically nonexistent.

In general, the role of dreaming in sleepwalking and sleep terrors, and the potential for dream interventions in these disorders, is as yet unknown. In adults, we know that sleep terrors and milder confusional arousals are more common in patients with mood disorders, which suggests some link to emotion regulation and mental health. In mild confusional arousals, a person partially awakens from NREM sleep and displays automatic movements (such as playing with bedsheets) along with moaning or unintelligible vocalisations. This can progress to thrashing about in bed or crying loudly, and the person may be slow to respond and difficult to wake up. We know next to nothing about these patients' dreams. Are these behaviours related to underlying dream content, and possibly repetitive nightmare scripts? Could they respond to behavioural dream interventions or nightmare treatments?

These questions are all the more important because current treatments for NREM parasomnias are insufficient. The basic clinical suggestion is to practise good sleep hygiene (follow a consistent sleep schedule, get enough sleep). To decrease the frequency of sleep terrors, parents can try waking their child up just prior to the time when episodes typically occur. In adults, sedative drugs can be prescribed in severe cases. Psychological treatments are lacking, although hypnosis and relaxation seem to be most helpful so far. But whether dream-related interventions could be useful in these disorders remains to be seen.

Moving away from the parasomnias, one of the most prevalent sleep problems in modern society is insomnia. About 30 per cent of adults report struggling to fall asleep or stay asleep, waking during the night or too early in the morning. While insomnia has mainly been studied objectively (using polysomnography or behavioural measures like time spent in bed), dream science has proven increasingly relevant to our understanding of this sleep disorder. Firstly, nightmares are very common in insomnia, as are dreams that are more 'thought-like' than typical dreams. Some patients with insomnia *feel* as if they are awake – thinking, worrying and restless – even when objectively asleep throughout the night. This is called 'sleep misperception', that is, misperceiving sleep as wakefulness.

Dream researcher Francesca Siclari took an interest in this phenomenon and revealed something interesting using high-density EEG. Even though typical measures used in sleep clinics (usually six scalp electrodes) might show the person to be asleep, higher-resolution imaging (with 256 electrodes) revealed areas of more wake-like activity in the brain, which corresponded with these patients *feeling* more awake. In essence, they dream of being awake, they dream of thinking and worrying during the night, and are unable to distinguish this sleeping experience from wakefulness. Most patients with insomnia have both objectively disturbed sleep and more sleep misperception. In other words, not only do they sleep poorly – they take a long time to fall asleep or wake up a lot during the night – but they also feel more awake and restless even when they do manage to sleep.

Even healthy sleepers sometimes experience sleep misperception – they report feeling awake in up to 10 per cent of early-night sleep episodes. This number is much higher in those with insomnia, around 40 per cent. In general, it's more common to feel awake during NREM sleep, and much less common in REM sleep – when subjects have more elaborate and immersive dreams. So although NREM sleep has long been thought of as the 'deeper' sleep, we now think that REM sleep and dreaming contribute more to the actual *feeling* of sleeping deeply. And for patients with insomnia, NREM sleep can even feel like wakefulness.

People with insomnia also feel more awake in the presleep period, thinking and worrying and delaying sleep onset. Most of us have amnesia for the minutes before sleep onset, meaning we forget our experience of falling asleep, but those with insomnia have *less* of this amnesia: they remember more of their experience falling asleep, and so it feels like they were awake even longer than it should. In other words, their mind is more active and awake, both before and during a night of sleep.

Bad Dreams and Health

Nightmares can also be a cause of insomnia, and many patients self-report that bad dreams are a major factor in their sleep quality. Along these lines, there is evidence that dream therapy can improve insomnia symptoms. For instance, a month-long trial of lucid dream therapy led to improved sleep, anxiety and depressive symptoms in fifty patients with insomnia. There are several possible reasons for this effect. First, lucid dreaming could help patients to become more aware of when they are dreaming, and by extension more aware of when they are sleeping (rather than mistaking sleep for wakefulness). Lucid dream therapy also increases the recall of more positive dreams, which corresponds with better sleep quality. And besides affecting dreams directly, patients practise presleep techniques like intention setting and visualisation, which could decrease worry before bedtime and improve sleep in a manner similar to other presleep mindfulness and relaxation techniques. Regardless of how it works, the use of lucid dreaming and other dream therapies hold promise for soothing the mind of insomnia patients before and during the night and improving their overall sleep experience and sleep perception, too.

To finish with one final sleep disorder: Sleep apnoea is arguably the most well-studied sleep disorder, characterised by repeated obstruction of the airway during sleep, and associated with numerous adverse health outcomes, such as cardiovascular disease. Sleep apnoea increases in prevalence with age, occurring in 15 to 25 per cent of adults over sixty.

While it is normal for breathing to decrease a bit during sleep, in sleep apnoea airflow either completely stops or reduces to only 10 per cent of normal levels, for ten seconds or more, dangerously reducing oxygen in the brain. Apnoeas can occur during either NREM or REM sleep, and patients with REM apnoeas have more frequent nightmares. Some scientists think that these nightmares result from the increased arousal and stress in the body following apnoeas. In support of this, one experimental study found that blocking a sleeper's airflow with a cloth induced nightmares. More generally, patients with sleep apnoea report more negative, violent and hostile dreams, although this is not the case for patients with *severe* sleep apnoea, who wake up so much that they have hardly any REM sleep at all. Patients with severe sleep apnoea, then, have suppressed REM sleep, and have fewer dreams and nightmares overall.

What's remarkable is that treating sleep apnoea with a CPAP machine (a machine that maintains airflow using continuous positive airway pressure) can lead to the disappearance of nightmares. For instance, 91 per cent of patients using a CPAP machine had reduced nightmares after use. Just one night of CPAP use can decrease unpleasant dream recall, even while increasing the actual duration of REM sleep. Importantly, for patients who have both sleep apnoea and nightmares, CPAP is the best first line of treatment for reducing apnoeas *and* nightmares. Said another way, traditional nightmare therapies are actually *not* helpful in cases where nightmares are due to underlying sleep apnoea. In fact, it is not uncommon for PTSD patients to struggle with both sleep apnoea and nightmares, and even in these patients sleep apnoea should be treated first.

*

This brings to light a final point: nightmares are not a homogenous sleep disorder. Nightmares can coincide with sleep apnoea or sleep paralysis, PTSD or sleep terrors – but their contents and causes vary across these conditions, and so, too, should their treatment.

As another example, little is known about the condition of 'epic dreaming', where patients complain that their dreams are too vivid and unrelenting across the night. Patients often have the impression of dreaming continuously of trivial, banal, or even physically demanding activities – like repetitive housework, or endlessly trudging through snow. These dreams can occur almost every night, continue after waking and returning to sleep, and are associated with daytime fatigue. Though epic dreams can lack emotion, they still produce waking distress, and their vividness and repetitiveness bear some resemblance to nightmares. Whether rescripting therapy or other dream treatments could offer any solace to these patients is as yet unknown.

In circadian rhythm disorders we know next to nothing about dreaming. For instance, in non-twenty-four-hour sleep-wake disorder, the body is not in sync with the twenty-four-hour day. This is common in blind individuals: in the absence of light cues signalling night and day, the body follows an internal clock set to slightly more than twenty-four hours. This means that their sleep-wake schedule shifts a bit more each day, becoming completely desynchronised with night and day over time. In shift work disorder, patients also have inconsistent sleep schedules due to shift work. While little is known about dreaming in these conditions, at least one intervention has adapted nightmare therapy to patients with shift work, addressing their unusual sleep schedules alongside imagery rescripting, and this resulted in fewer nightmares, less daytime sleepiness, and decreased insomnia, depression and PTSD symptoms.

Targeted approaches like this that account for the specific sleep and dream experiences of different populations could be a valuable next step in sleep medicine. While most sleep disorders have been studied primarily through a behavioural or biological lens, we have increasing evidence that dreaming is relevant to understanding and treating these conditions, too. Overall, further study of the role of dreaming, and the therapeutic potential of dream-related interventions, is needed to pave the way towards a more holistic field of sleep medicine.

Dreams and mental health

The past few decades of sleep and psychiatric research have repeatedly confirmed the importance of adequate and quality sleep for good mental health, although still scant attention is paid to dreaming. In truth, there is immense stigma attached to the word 'dreaming', likely a backlash from the dominant era of psychoanalysis. This is so much the case that many of us dream researchers skirt around the term in different scientific or medical venues, opting instead for phrases like 'subjective experience', 'sleep perception', 'consciousness', or we rely on the more reputable 'nightmare' to wedge our foot in the door of medicine.

Even nightmares, though, are almost never assessed or treated in psychiatry, although my perception is that there is growing awareness of their importance. Working in a psychiatry department over the past few years, I have heard from numerous colleagues about how many of their patients complain of nightmares, and that they would like to learn how to diagnose and treat nightmares in their practice. Most clinicians do not yet have the tools to help patients,[3] given that there is a lack of sleep and nightmare training in medical school. Such training would be highly valuable, since nightmares are linked to all manner of mental health disorders and to the severity of symptoms in these disorders, too.

As we've learned, nightmares are especially pervasive in PTSD, and they are also linked to anxiety, depression and suicide risk. Over the past decade, new evidence has shown that nightmares contribute to many other psychiatric disorders, including psychosis and schizophrenia, borderline personality disorder and even addiction and eating disorders, too.

To start with the importance of nightmares in psychosis: over half of psychosis patients have nightmares, and the occurrence of nightmares correlates with the severity of waking symptoms and with their fluctuation over time. Nightmares can actually predict spikes in symptoms: first, anxious and apprehensive nightmares often precede the initial onset of psychotic symptoms, and the recurrence of nightmares can predict further bouts of psychosis later on. In the short term, on nights when patients have more negative dreams, they also have worse paranoid and hallucinatory symptoms the next day. That psychosis can be, to some extent, predicted by the presence of certain kinds of dream content is astonishing – and the fact that this facet of a person's life is not currently used in diagnosing or treating most mental health conditions is a stark reminder of how far medicine has to go.

A small pilot study explored whether treating nightmares in psychosis is feasible, with a modified form of imagery rescripting therapy. Because patients with psychosis often do not respond well to exposure, the first step of exploring and feeling the negative emotions in a nightmare was limited; instead, patients only briefly described their nightmare before moving on to rescripting. They were instructed to devise an alternate ending that could be inserted before the peak of the nightmare and to resolve the distressing theme of the nightmare in a positive or empowering way. Patients were then instructed to close their eyes and re-enter the dream in vivid sensory detail, re-experiencing the dream in first person and present tense through to its new ending, and to hold on to the positive emotion at the end of the sequence. After rehearsing this positive imagery on a daily basis for six weeks, all five patients saw changes in their nightmares and felt less distressed by them, though the nightmares remained frequent. The patients reported feeling more able to cope with and control nightmares, no longer waking up in terror, and even visualising pleasant imagery after awakening from a nightmare. One patient described the shift in their nightmares as feeling like they had 'deconstructed the memory and put it back together again ... It's less dark, both in terms of how I feel and what I see in the dream.'[4]

Across other psychiatric disorders, too, treating and paying attention to nightmare patterns can have substantial impacts. To highlight again the importance of nightmares in suicide: research has now revealed a pattern of changes in dream content that precede a suicidal crisis. In patients hospitalised for suicidal crisis, an uptick in bad dreams started around four months prior to hospitalisation, then more severe nightmares three months prior, and finally depictions of suicide and death appeared in dreams in the final month before hospitalisation.[5] Depictions of suicide and death in dreams could reflect increasing suicidal thoughts in waking life and even spur on suicidal thoughts after their occurrence. That these dreams offer a clear indication of a patient's risk has led some clinicians to argue that asking patients about their dreams should be a mandatory and potentially lifesaving component of suicide risk assessments. In attempts to understand what it is about nightmares that could be so strongly related to suicide (more so than other measures of depression, insomnia, or anxiety, for example), some research points to feelings of hopelessness and impulsivity that result from nightmares; patients lose any hope in the possibility for change or control over their life or their nightmares, and they have reduced inhibitions and poor emotion regulation due to sleep loss. Screening and treating nightmares is a simple and significant way to improve prevention of suicide. And indeed, treating nightmares in psychiatric patients reduces their suicidal ideation in the long term.

Beyond suicide risk, nightmares are also prevalent in borderline personality disorder (BPD), and patients with more distressing nightmares have worse dissociative symptoms, more self-harm and substance abuse behaviours, and more repeated suicide attempts than those without nightmares. The occurrence of nightmares seems closely tied to patterns of rumination and negative thought spiralling during the day, which is common in BPD. Since we know that rumination and worry prior to sleep is a trigger for nightmares, positive imagery rehearsal in the presleep state could help to reduce nightmares and lessen negative thoughts and emotions in the days following. A clinical trial in 2024 was the first to explore whether treating nightmares is beneficial in this population. In twenty-two patients with BPD, imagery rehearsal therapy led to reduced nightmares and improved waking symptoms compared to patients undergoing treatment as usual. Those in nightmare therapy also had less intrusive thoughts and lower arousal and anxiety at the end of therapy.

It is worth repeating that in many psychiatric disorders, it appears to be the case that emotion dysregulation is key to the downward spiral between nightmares and psychiatric symptoms. This aligns with the model of nightmares discussed earlier, where poor emotion regulation is central not only to the generation of nightmares but also to the negative consequences of nightmares on waking mental health. By treating nightmares directly, both sleep and emotion regulation can be repaired, which improves mental health thereafter. Because of this, it is highly likely that treating nightmares could benefit many different psychiatric patients, including those with psychosis, depression, bipolar disorder and borderline personality disorder – all conditions where nightmares are currently undertreated and where emotion dysregulation is a core component.

Patterns in disturbed dreaming can also be useful in diagnosis, differentiating patients with different disorders.

Using what's called a speech graph, scientists can visualise patterns of speech, such as how often patients repeat certain words or concepts and the diversity of words used in a report. Patients with different psychiatric disorders use measurably different speech patterns when reporting their dreams, compared to healthy subjects, and even when compared to each other. As an example, patients with obsessive-compulsive disorder (who suffer from intrusive thoughts and urges to perform repetitive behaviours) tend to repeat the same words in dream reports and use a less diverse set of words than healthy subjects. Speech graphs of dream reports, specifically, are better able to distinguish among patients with psychosis, OCD, bipolar disorder, or no disorder, than other types of verbal reports. To explain what I mean by that: when subjects complete several different types of reports – including a verbal description of a dream, a recent or old memory, and a positive or negative image – only speech graphs of dream reports are able to reveal patient diagnosis, suggesting that dreams are somehow more sensitive to altered thought patterns in these disorders. It's possible that because dreams are less tethered to the more stable (and mutually agreed upon) external world, they are more telling of a patient's inner thought patterns and memory organisation.[6]

In related work, other measures of dream content also vary between different patient populations. For instance, patients with schizophrenia generally have dampened dream recall and shorter dream reports, with fewer emotions overall. Some studies have found their dreams to feature more strangers and fewer familiar characters or friends, with more hostility from other characters in their dreams. The perceived hostility from strangers and unfamiliar characters is consistent with symptoms of persecutory delusions in waking life. And although patients with schizophrenia do not judge their dreams to be more bizarre than the dreams of healthy subjects, external judges rate their dreams as more bizarre and implausible, perhaps reflecting symptoms of disorganised thinking.

These specific features of dream content differ from those of other mental health conditions, such as depression, where patients have nightmares marked by failures and misfortunes. Patients with depression also play a relatively passive role in many of their dreams and, besides nightmares, experience emotionally flat and even colourless dreams that seem to reflect patterns of emotional numbing. At the same time, nightmares with increasing references to death can reveal an uptick in suicidal thoughts. In patients with bipolar disorder, shifts in dream emotions can predict changes in waking mood: going from more anxious to positive and bizarre or unrealistic dreams can indicate the onset of a manic episode.

Altogether, this kind of research is revealing how dreams can be uniquely informative to mental health status and can potentially help clinicians to diagnose, follow and treat patients over the course of disease. In particular, being aware of changing patterns in dream content can alert both a patient and a clinician to oncoming symptoms.

In fact, that dream content can act as a signal to clinicians is perhaps most evident in the case of addiction.

A full 80 per cent of patients with addiction report 'drug dreams' in the first few months of abstinence – where the dreamer attempts to use or seek out their drug of choice within a dream. Most research finds that drug dreams are strongly tied to waking cravings – even increasing cravings the next day. This seems especially true for dreams where the dreamer is endlessly searching for drugs but is unsuccessful in obtaining them, and then wakes up in an intense state of craving; or dreams where drug use is pleasurable, and the patient wakes up disappointed in the morning. These dreams occur in all forms of addiction – alcohol, heroin, cocaine, tobacco and benzodiazepine use – and occur especially early in recovery. Behavioural addictions, too, are associated with these kinds of dreams. In pathological gambling disorder, patients report 'gambling dreams' where they are playing games or seeking a big win, or witnessing others gambling and feeling an urge to join in.

These addiction dreams can be a valuable tool for clinicians, acting as a sort of barometer and alerting to oncoming periods of craving. A sudden appearance of drug dreams could indicate a risk of relapse and prompt a therapist to reinforce strategies for managing cravings and preventing relapse. At the same time, some drug dreams seem to reaffirm sobriety, namely, if there is resistance to use in the dream and relief on awakening. To give an example, a patient might dream of having a drink to celebrate one month of sobriety, and after taking a shot suddenly feel horrified, realising all progress is lost. Dreamers often feel disbelief and guilt that a relapse has occurred while within the dream, and then feel relieved on awakening, grateful it was just a dream and more determined to stay abstinent. In this case, the recurrence of these dreams could actually protect against relapse over time.

It's worth noting that drug dreams decline on their own after a period of sustained abstinence (as we saw earlier, it takes some time for the dreaming brain to reorganise). Drug dreams are more prominent in early sobriety when a patient is actively trying to stop their use, maybe even suppressing a pathological urge all day, which then rebounds into dreams at night. While drug dreams generally decrease over time, they can still recur during periods of stress or transition, similar to other types of recurring bad dreams. This can be frustrating for patients, who may worry that these recurring drug dreams mean they will never fully recover.

On the one hand, simply learning that drug dreams are a normal part of recovery, and are not a reflection of failure or lack of motivation, can be reassuring to patients. There is also some evidence that working with dreams in a group setting can be helpful for patients recovering from addiction. Moreover, if drug dreams are considered a variant of nightmares, it's possible that rescripting therapies could offer a means of alleviating these recurring bad dreams, potentially reducing relapse risk and craving over the course of recovery. While attempts to treat drug dreams via nightmare therapy have not been studied, there have been case reports of pharmacological treatment of drug dreams with prazosin, which is a high blood pressure medication that has been used off-label to treat nightmares. For instance, one patient who had distressing dreams of injecting methamphetamine night after night in early abstinence was prescribed prazosin and immediately these dreams disappeared.[7] The same was true for a patient with nightly cocaine-using dreams. Prazosin seems to work by reducing arousal during sleep and thereby suppressing nightmares (though this is not fully understood), but unfortunately, if prazosin use is discontinued, nightmares often return. So while useful in early recovery, in the long term, patients would likely benefit from psychological treatment of nightmares.

Somewhat similar patterns to drug dreams emerge in patients with eating disorders. These are patients who have an intense preoccupation with food during the day and are constantly trying to avoid or restrict their eating behaviours in waking life. This leads to an increased prevalence of food and eating themes in dreams. In general, patients with disordered eating also have more bad dreams and nightmares, and bad dreams can trigger episodes of binge eating in patients with bulimia, similar to how drug dreams intensify waking cravings. I remember one of my subjects from a longitudinal study in the UK who dreamed of eating delicious foods on a near-nightly basis. My immediate suspicions were later confirmed when this subject disclosed how her dreams were related to a history of disordered eating. Though 'eating delicious foods' is a fairly typical dream theme, it is rare for it to occur with such frequency in an individual. More generally this highlights how personal our dreaming patterns become, and how they can reveal our waking concerns and even at times divulge our otherwise concealed mental health struggles.

In general, monitoring dreams can act as a window into mental health, and more attention to them in clinical care is warranted. Nightmare therapies can be easily added on to other waking interventions and targeted treatment would alleviate not only nightmares but also waking symptoms, thanks to restored sleep and repaired emotion regulation, too. Parallel work by one of my mentors, Wilfred Pigeon, has shown time and again that treating insomnia in numerous psychiatric populations (depression, PTSD, chronic pain) leads to better outcomes than treatment for waking symptoms alone. When it comes to nightmares, their treatment offers a powerful first-line defence against suicide, an immediate access to healthier emotion regulation and even lasting protection against future mental illness in certain at-risk populations.

Dreams and physical health

Beyond sleep and mental health, there is increasingly undeniable evidence of dreaming's relevance to physical health: during pregnancy, after physical trauma, in autoimmune disease and even at the end of life.

In the first place, we know that dreaming goes through several marked changes over the developmental lifespan, and one particularly notable case is during pregnancy. Pregnant women have twice as many bad dreams and nightmares as non-pregnant women, and 80 per cent of new mothers report that their dreams took on a particularly vivid and bizarre form during pregnancy. Up to a third of pregnant women's dreams relate to pregnancy, and these dreams can feature terrifying images of childbirth or disturbing threats to the baby's or mother's health. Even though these dreams are common, their morbid content can be unnerving to unsuspecting new mothers. Frequent nightmares and baby-themed dreams could be due to several factors, including of course increased psychological stress, but also increased nighttime awakenings, and even physical sensations of foetal movements interrupting sleep and dreams.

After childbirth, new mothers continue to experience bizarre bad dreams and nightmares, which often feature concerns about motherhood and parenting. One common theme is the 'baby in bed' dream, where around 50 per cent of postpartum mothers (and some fathers) dream that their newborns are lost in the sheets or have fallen underneath the bed and are in danger of suffocating or are even dead. Many women start frantically searching for their infant while still asleep, groping around the blankets or grabbing their partner's face, speaking or crying and in a confused state on awakening. Even upon waking and realising the baby is not in bed, the realism of the dream compels most mothers to get up and check on their infant.

Also prior to pregnancy, many women report disturbing baby-themed dreams, with bizarre content like realising one has had a baby but forgotten about it for some time and neglected to feed it, or giving birth to a baby that is abnormally small, frail, or unusual. Since many women are not aware of these phenomena, simply learning about these naturally occurring dreams can help to lessen the distress associated with them, understanding that these dreams are not a cause for concern. While as yet untested, it is also possible that nightmare therapy could reduce the occurrence of bizarre nightmares during pregnancy or lessen baby-in-bed dreams after childbirth. Instilling more positive dreams around pregnancy, birth and motherhood could also reassure and support mothers in this new phase of life.

On the other end of the life cycle, the dreams of dying patients are also unique, and uniquely impactful. Up to 87 per cent of dying patients report profound end-of-life dreams that provide reassurance or guidance in the transition to death and lessen the fear of dying. These end-of-life dreams can occur in the days and weeks before death, and have been documented throughout history and across cultures and religions. End-of-life dreams are highly realistic, and dreamers often relive meaningful memories, reuniting with lost loved ones, or preparing to go somewhere beyond this life. As an example, in the days before one woman died, she reported dreams of seeing her deceased husband waiting for her at end of a staircase.[8] These dreams can provide personal and spiritual solace, helping patients reconcile with their past and accept approaching death. Family members and caregivers even report that patients who have such dreams experience a more peaceful and calm transition to death.

These end-of-life dreams are in stark contrast to the nightmares reported by some patients struggling to survive in intensive care units (ICUs). ICU patients experience vivid nightmares that are filled with horror or dread and often portray painful impending death. These dreams can depict a patient's real experience in the ICU, sometimes marked by painful treatments, isolation and the constant fear of death. Several studies attest to the high prevalence of ICU nightmares and even describe them as having a traumatising long-term impact on patients. In fact, nightmares are the most frequently remembered traumatic experience reported by ICU patients (in 64 per cent of patients). They are described as 'bizarre and extremely terrifying' and contribute to the later diagnosis of PTSD.[9] A condition called ICU delirium can also occur (in 37 per cent of cases), which includes a set of symptoms: nightmares, disorganised thinking and disorientation, paranoid delusions and hallucinations. Stressful conditions contribute to both delirium and nightmares in the ICU, such as being intubated or under sedation, experiencing pain, noise, or discomfort. Contrary to end-of-life dreams that evoke peacefulness and acceptance of death, ICU nightmares evoke intense fear and portray a patient's struggle to survive. Attempts to improve conditions in the ICU to reduce delirium and nightmares are now being studied, especially because nightmares and delirium have been associated with worse outcomes for these patients even after leaving the ICU.

In a related context, in hospice care, dream sharing has been explored as an intervention to support patients who struggle with a loss of meaning at the end of life. In one study, patients discussed their dreams with a group and together examined the meanings of their dreams and how they related to their lives. These dreams often featured family and friends, and familiar settings and feelings. Some patients felt they needed to change their way of thinking before dying or complete unfinished business. Over the course of twelve sessions, patients described dream sharing as meaningful and comforting, leading the researchers to suggest this as a supportive form of counselling for the dying.[10] We'll see later how dream sharing can also provide support to those living with terminal illness or suffering from bereavement, periods of life marked by confrontations with death, when dreaming and dream sharing can provide comfort and clarity.

*

So far, then, dreaming is clearly altered during major physical transitions in life, and before death. What about in other physical health conditions? Could bizarre nightmares predict seizures or migraines, or flare-ups of arthritis or chronic pain? Some studies indicate yes, but what could explain such findings?

To start with, in patients with chronic pain, waking aches and discomfort often seem to intrude into their sleep and dreams. About a third of patients with chronic pain report feeling pain in their dreams, which is much higher than the 1 per cent of dreams containing pain in healthy people. The amount of pain experienced in dreams correlates with the severity of pain symptoms in waking life, and, more generally, patients with chronic pain have more negative and aggressive dreams and nightmares overall. It could be that physical sensations of pain are being incorporated into dreams. For instance, applying extreme pressure to the leg increased the incidence of pain and negativity in dreams in an experimental study, like one subject who dreamed: 'I could barely make it up the stairs and I think I was dragging my leg ... because it was hurting. I would reach down and grab my leg and pull it along with me. It felt really heavy, like it was a wooden leg.'[11]

From another angle, pain and negativity in dreaming could be related to the psychological distress of dealing with chronic pain in waking life. One case report described the resolution of chronic pain after a transcendent lucid dream, in a patient who had suffered for forty years with unbearable and untreatable symptoms. The patient dreamed of hearing *tones so pure and composed in such a wonderful way... As the tone kept playing, I remember it was so euphorically beautiful ... extraordinary. It was like a near-death experience ... a wonderful vision of clarity and beauty and composition. It was like my brain had shut down and been rebooted.*'[12] Field studies have also supported the use of healing lucid dreams for chronic pain relief, while other studies have explored the use of waking imagery rescripting to reduce chronic pain. For instance, in an intervention similar to exposure and rescripting, patients were guided into a state of relaxation and awareness, and instructed to develop sensory imagery where they could physically feel and then alter their pain. After four days of this practice, patients had reduced pain compared to a control group. Variations of guided imagery have been shown to alleviate pain in patients with fibromyalgia, or those with malignant tumours, and motor imagery has also proven beneficial for patients with chronic low back pain. Finally, more generally, we know that poor sleep can worsen pain symptoms, and that treating sleep improves pain symptoms. While novel, these findings suggest that modifying psychological components of chronic pain through imagery work, and treating sleep directly, can benefit waking symptoms of pain.

In still other cases, physical diseases seem to protrude into dreams even before their diagnosis in waking life. One study of patients with near-fatal cardiac events looked at the incidence of so-called 'killer dreams', dreams of death or pain in the arm, heart, chest, or neck that precede a cardiac event. One patient dreamed he was murdered, and awoke with crushing chest pain prior to cardiac arrest an hour later.[13] Other dreams appear to precede oncoming epileptic seizures or migraine attacks, including dreams of being shot or struck by lightning in the head, which patients claim can alert them to oncoming episodes. Anecdotal evidence is replete with stories of these kinds of dreams that predict oncoming health problems, known as prodromal dreams. Case reports in medicine include examples of patients dreaming of the presence or location of a tumour prior to being diagnosed with cancer, or dreaming of wounds or hearing prescriptive instructions before or at the onset of disease. It is possible that subtle physical symptoms are detected and incorporated into dreaming early in disease onset, and prior to more significant waking symptoms. More conservative views regard prodromal dreams as coincidence or hindsight bias, where patients retrospectively identify a dream as predicting later health.

At least one area of physical medicine suggests that disturbing dreams and nightmares really do predict upcoming surges in physical symptoms. In autoimmune disorders, very high rates of nightmares occur and often precede flare-ups, as a sort of prodromal symptom. In neuropsychiatric lupus (a kind of lupus that affects the brain and nerves), 61 per cent of patients report increasingly disturbed sleep and vivid distressing nightmares before flare-ups, which often feature physical threats such as falling or being attacked, trapped, or crushed, or even car or plane crashes. In rheumatoid arthritis, 30 per cent of patients have weekly nightmares, over five times higher than in the general population, and patients also report frequent daymares – vivid, uncontrollable and frightening daydreams – that increase in occurrence before flare-ups. Recognition of these nightmare and daymare patterns could help patients and clinicians to identify an impending flare-up, initiate treatment and potentially avert further symptoms at any early stage. Nevertheless, physicians to date seem sceptical of the value of asking about nightmares in routine care, preferring to focus on physical symptoms alone.

This is not surprising, given that empirical study of prodromal dreams and even nightmares in physical medicine is still scarce. And yet, it is clearly the case that physical changes associated with pain and disease, and developmental changes during pregnancy and at the end of life, are reflected in unique patterns of dreaming. These patterns remind us that dreams have their foundations in our physical bodies and are fundamentally rooted in our physical health. It is perhaps time that a paradigm shift is called for, acknowledging that how we experience sleep is inextricably tied to our well-being, and offers a valuable source of information in all areas of medicine.

All in all, a renewed and growing interest in dreaming across medical spheres is confirming that how we experience this full third of our lives is both meaningful and consequential to our health.

We know that bad dreams often crop up during periods of psychological stress, but they also occur in tandem with changes in sleep patterns or physical health. All manner of dreams, from bizarre and disturbing to peaceful and realistic, can serve as useful markers and provide clues to progression or recovery over the course of a disease, such as in Parkinson's, addiction, depression and more. Other health-related dreams seem based in physical sensations, like pain, drawing on a fundamental property of dreaming, that dreams are built into our bodies. We know that nightmares aggravate symptoms in all manner of disorders, including PTSD and psychosis, chronic pain and insomnia, and, if left untreated, lead to worse outcomes in these illnesses, too. In psychiatric conditions at least, poor emotion regulation seems to be key to this downward spiral between distressing nightmares and worsening waking symptoms.

Given all of these links, treating nightmares directly can dramatically improve prognosis, but nightmare treatments, while effective, are still rarely available to patients.

The first major hurdle to treating nightmares is that even sleep clinicians, along with psychiatrists and physicians, largely do not ask about nightmares. If nightmares do come up as a problem, a common response is to prescribe medication to suppress REM sleep, and therefore suppress dream and nightmare recall altogether. Unfortunately, upon stopping medication, patients often experience what's called a REM rebound, a dramatic increase in REM sleep along with disturbing dreams and nightmares. Moreover, it's unclear whether prolonged REM suppression could damage sleep and dream function in the long term.

To give some examples: clonazepam is a drug primarily used to treat seizures, but it is also prescribed for RBD to reduce dream enactments and disturbing dreams; while effective in the short term, cessation is followed by a REM rebound, a resurgence of nightmares and dream-enacting behaviours. Medication has also been prescribed for nightmares in addiction, PTSD and bipolar disorder, namely prazosin, which is an antihypertensive medication that reduces blood pressure and arousal. This medication does effectively reduce nightmares, but it also generally suppresses REM sleep and can lead to a rebound of nightmares later on. Most antidepressant medications also reduce REM sleep, and while these do initially improve mood, the long-term effects of REM sleep (and dream) suppression on emotional well-being are not well understood. Even sleep medications often work by suppressing REM sleep, though this is rarely noted as a side effect. If REM sleep and dreaming are functional in the many ways described throughout this book, then these medications may not be the best long-term solution for health conditions.

Instead, as laid out in this and prior chapters, working *with* dreams and nightmares, rather than avoiding or suppressing them, could be our best approach to healing sleep in many cases. Paying attention to dreams, as we're beginning to uncover, can help clinicians to better diagnose, care for, and treat their patients. And on a more positive note, what we'll see next is that we don't need to wait for things to go wrong to tap into the usefulness of dreaming. Dreams have all sorts of potential for supporting our well-being: enhancing physical learning, creativity, or even community-building. With simple tools of dreamwork and dream engineering, these resources for a healthier dreaming and waking life become available from the moment your head hits the pillow each night.

8

Sleep On It: Dream Skills

So far, we've seen all the ways dreaming and especially nightmares are related to health, from sleep medicine to psychiatry and all phases of physical development or disease. And we've seen how dreaming can act as a barometer, revealing recovery or relapse in addiction or suicide risk; or as a portal for intervention, such as treating nightmares in narcolepsy, or even improving sleep quality in insomnia with lucid dream therapy.

We have also explored the one-to-one work that my colleagues and I do with nightmare sufferers and how dreams and nightmares are constructed and experienced in the lab. We saw the typical designs of dreams, the scaffolds of the dream world and potential functions of dreaming – for emotion regulation or skill rehearsal, for example. In this chapter we'll return to the concept of skill rehearsal, to learn more about how different skills can be harnessed through practices of dreaming, dreamwork and dream engineering. These 'dream skills' can support our health and well-being in waking life and, importantly, be utilised by anyone.

One of the most frequent comments I receive after giving public talks is about the creative potential of dreams. I've had people come up to me to relate an impressive anecdote about a roommate who could take a nap and instantly resolve any creative problem, or a family friend who is a mathematician and regularly solves theorems in his nightly dreams. They wonder how this is possible – and whether it's a skill that they, too, can hope to attain. As we learned earlier, REM sleep and dreaming are thought to occur in a hyper-associative state, where new links in memory are unearthed and recombined, giving rise to unusually creative ideas and insights. While evidence before has been largely anecdotal, science is now revealing that the creative power of sleep and dreams is readily available to those who know how to unlock its potential, and we'll learn how to in this chapter.

Moreover, dreaming is linked to all kinds of learning, and one way that dreaming offers a unique space for learning is that our experience is not just imagined, it is embodied. Experience, after all, is the best teacher, and in our dream worlds, we get to relive all sorts of physical memories in freely-moving dream bodies and multisensory worlds. In this sense, dreaming can especially support sensorimotor skills. There are numerous anecdotes and studies of elite athletes or musicians who use lucid dreaming to enhance their performance. One example is of a swimmer who does laps in a pool full of jelly (in his dream) to train his body to swim against greater resistance and slow down the flow of each stroke. It seems like a form of implicit learning unfolds within the dream, and a renewed sense of muscle memory is found available within the body in the morning, similar to the sudden appearance of creative insight in the mind.

While the idea of becoming an athlete overnight is appealing, this implicit learning stretches beyond physical skills, into the social realm too; dreaming, after all, has a fairly reliable social aura to it – how we are always meeting friends or strangers in the sex dreams and ex-dreams of our nightly lives. We've all had that bad dream about our partner cheating or leaving, dreams that leave a bad feeling in our gut in the morning. Do the social lives of our dreaming mind matter? I know this is a concern for many, as my most-read blog post is sought by dreamers who are wondering 'What does it mean when I have a bad dream about my partner?' As it turns out, our densely populated dream worlds help our hardwired social brains to strengthen social skills and maintain a sense of connection while we sleep. We can also learn how to use dreams as a form of social support – in times of grief, loneliness, or social isolation.

What are the limits of these dream skills? Just how far can we go with the phrase 'sleep on it'? And where does science meet fiction? Can we, like in *Inception*, plant ideas into the unconscious mind to stir the pot of creative insight each morning? Can we harness dreaming for physical rehabilitation, use this embodied rehearsal space to control an avatar with the sleeping mind? And where do past lovers dwell in our dreams? Can we extinguish heartbreak, to bring eternal sunshine to our waking lives? Are the possibilities endless, or limited, and what does the future hold?

Sleep On It: Dream Skills

Inception: dreams and creative insight

Dreams have inspired numerous artists and inventors throughout history, with well-known examples like the melody for the song 'Yesterday' arriving one morning in Paul McCartney's inner ear as a dream-tune, and the structure of the periodic table of elements famously dreamed up by scientist Dmitri Mendeleev. Author Anne Rice's dream about a petrifying woman made of marble inspired the plot of *The Queen of the Damned*, and Robert Louis Stevenson claimed that 'little people' in his dreams were to thank for Dr Jekyll and Mr Hyde.[1]

Where do these ideas come from, and what is it about dreaming that lends itself so readily to creativity?

As we learned earlier, our dreams seem to link together foreign thoughts and concepts into new and unusual, creative combinations. Our (REM) sleeping brains search untravelled paths in memory, instead of dredging the usual paths taken in wakefulness. Remember that REM sleep increases access to distantly related memories, especially useful for creative insight, allowing us to find those uncommon ideas inaccessible to the logical mind. In fact, when faced with a problem with no apparent solution, it is often helpful to sleep on it, after which the solution might appear in a burst of insight. A spate of sleep works even better than staying awake and focusing on the task, presumably because our waking thoughts often get stuck on obvious and unoriginal solutions, failing to find the hidden key.

Already science has indeed proven that 'sleeping on it' leads to creative insight, but is this a product of sleep alone, or do dreams themselves spark creativity, too?

Of course, through numerous anecdotes and surveys we know that many people *believe* that their dreams, which are often original and bizarre, are to thank for sudden insights or the discovery of creative ideas. But despite these perceived links between dreams and problem-solving, there has been little experimental proof that dreaming begets creativity (that is, that dreaming boosts creativity beyond the benefits of sleep alone). The lack of evidence is certainly not due to a lack of interest in the topic; researchers have revealed that everything from dream recall to dream vividness is linked to higher creativity and imaginative skills (such as being able to reorient 3-D objects in the mind) and even to having a career in the arts. But these studies proved only that more creative people have more vivid and frequent dreams, and whether the content of dreams itself provides a source of creativity remained untested and unknown – until now.

Dream engineering has made it possible to test whether dreaming itself is to thank for creativity: by controlling and manipulating dream content, we can test how this impacts creative performance.

To give an example, targeted reactivation can be used to associate sensory cues with creative problems, riddles like 'What is the meaning of HIJKLMNO?' (the answer being water: H_2O). When cues are re-presented during sleep, subjects are better able to solve these problems later, after waking up, as insights seemingly rise to the surface from their unconscious. To examine the role of dreams in this process, a team of researchers at Northwestern University trained sleepers to become lucid and instructed them to intentionally work on the puzzles in their dreams.[2] After the dreamers attempted to solve several puzzles prior to sleep, each associated with a unique sound, some of these sounds were re-presented during REM sleep that night, to remind dreamers to work on the unsolved problems in their dreams.

What the scientists discovered was that, first off, sensory cues produced more dreams about the unsolved puzzles, even when subjects were not lucid. Independent judges rated both non-lucid and lucid dreams as containing more references to the cued puzzles than to puzzles that had not been cued. Importantly, this targeted dreaming benefited problem-solving – those who were cued and dreamed of the puzzles were more than twice as likely to solve the problem come morning. Those who did not dream of the puzzles did not improve, regardless of whether they had been cued, suggesting the dreams were more important than sleep itself.

Several subjects also became lucid and gave signals to indicate they were working on the problem in their dream: first, a left-right eye signal to indicate they were lucid; second, a series of sniffing patterns (detected by a tube set below the nostrils) when they started on the puzzle. There were nineteen lucid dream attempts to solve puzzles in total, and most of these (74 per cent) led to success the next morning; these dreamers reported thinking about or visualising the puzzle to try to find solutions in their dream. In the few unsuccessful cases, the dreamers looked to other characters for help, who turned out to be quite useless in the end. One dreamer asked the experimenter in her dream: 'How do you do the four-shape puzzle?' to which the character responded 'Well, I actually don't know. It's kind of hard.' Believe it or not, this is consistent with a prior study that found that other dream characters were less able to solve puzzles in dreams than lucid dreamers themselves, so it seems dream characters are not the most logical bunch!

In general, though, assuming we're not being led astray by our supporting cast, targeted dreaming was able to elicit more dreams of unsolved problems, and this led to insights regarding these puzzles the next morning.

Given its creative potential, it's perhaps unsurprising that many writers and artists report using lucid dreaming to intentionally work on their craft, in essence harvesting these dreams towards creative pursuits.[3] Some report simply observing dreams for inspiration, whereas others actively bring their work into the dream, summoning a character from a novel and engaging in dialogue, or entering into a scene or plot point and watching as a new script unfolds. This can result in surprising and creative narratives, as the dream world emerges of its own volition, with characters who seemingly have their own thoughts and feelings and intentions. Moreover, it lends a sense of verisimilitude and immersion, spontaneity and novelty to the personas and places in story-writing. Creative writers especially report world-building within dreams, being able to touch and feel, see, hear and smell the scenery. Artists gain lived experience of their imaginary worlds, and an authenticity imbues the people and places visited in dreams, which later come to life on the page or canvas.

Of course, lucid dreams are known for being perceptually rich, elaborate and easy to recall, making them uniquely valuable for artistic inspiration. In this sense, visual artists also like to experiment with lucid dreams, like artist Dave Green in London, who generates drawings within lucid dreams that he re-creates on waking.[4] Following one of our conversations about how emotions shape dreaming, Green became lucid and started sketching while shouting angrily into the dream. As a result, several screaming figureheads appeared on a (dream) piece of paper. The best part was that when he stopped shouting and put the paper down, he noticed that all the heads in the drawing had stopped shouting too, saying: 'It was so cool in the dream to see my emotions reflected like that. It was like the drawing was shouting back at me.'

For artists, part of the wonder of lucid dreaming is in being able to witness such effortless creativity, a seemingly boundless flow of imaginative prowess emerging in the sleeping mind.

While lucid dreaming experts can tap into this resource repeatedly from one night to the next, or even iteratively across a single night, the fact is that most of us do not have such easy access to lucid dreams. Luckily, their more modest neighbour – the sleep onset microdream – is also surprisingly potent for creative insight, and much easier to access, too.

Salvador Dalí, a master of surrealist art, often lifted images from the sleep onset state to inspire later paintings. In his famous 'slumber with a key' technique, Dalí enters momentary naps 'less than a quarter of a second long' to access the fluid space between wake and sleep, where the myriad memories and sensations of the day collide into microdreams. Dalí performed this micronap whenever needed to obtain both visual inspiration and rest, refreshing his mind and body for the labours of painting.

> You must seat yourself in a bony armchair, preferably of Spanish style, with your head tilted back and resting on the stretched leather back. Your two hands must hang beyond the arms of the chair, to which your own must be soldered in a supineness of complete relaxation. [...]
>
> In this posture, you must hold a heavy key which you will keep suspended, delicately pressed between the extremities of the thumb and forefinger of your left hand. Under the key you will previously have placed a plate upside down on the floor... The moment the key drops from your fingers, you may be sure that the noise of its fall on the upside down plate will awaken you.[5]

In this procedure, the muscle paralysis that occurs upon falling asleep causes the key to drop and startle a person awake, with a hypnagogic image fresh in mind.

Dalí claimed to resolve the problem of 'sleeping without sleeping', gaining the benefits of sleep without a complete loss of wakefulness. This is quite accurate to modern science, which views stage 1 sleep as a sort of hybrid state with elements of both sleep and wake. As we learned earlier, sleep onset is a process: while parts of the brain disconnect from the outer world and start to fall asleep and tune into the wandering mind, other parts remain awake. Neuroscience studies confirm that creative thought requires a balance: between freely generating novel, if bizarre, ideas and consciously sifting to find which is useful. Dalí would lightly tread this tightrope, dipping into sleep's creative abundance, then promptly waking to catch useful images as they arise.

Stage 1 sleep offers the perfect mix between spontaneous, flexible thinking (being hyper-associative like REM sleep), and a proximity to the logical mind, which is able to recognise and grab creative ideas on the dime. In fact, spending only *one* minute in stage 1 sleep nearly tripled creative insight in one study,[6] where over a hundred subjects attempted to solve math problems that had a hidden solution. After first attempting (and failing) to solve these problems while awake, subjects enjoyed a twenty-minute siesta, relaxing in a semi-reclined chair with their eyes closed while being monitored by PSG. Similar to the Dalí method, they held an object in their right hand and if the object fell due to their momentarily falling asleep, they reported out loud their stream of thoughts from just before awakening.

The discovery of the hidden solution was 2.7 times more likely after only a minute of stage 1 sleep, compared to a period of wakefulness. This creative benefit was lost if participants slept too long or if they descended into stage 2 sleep. There seems to be an ephemeral period where insights are unlocked at sleep onset, what the researchers call a 'creative sweet spot' favouring the 'exploration and capture of remotely associated concepts'.

Is it simply that the brain shuffles thoughts into creative combos in this critical moment of sleep onset, or was Dalí right: do microdreams play their part in creative thinking, too?

On the one hand, it has been difficult to test whether dreams themselves benefit creativity, above and beyond the clear benefit of stage 1 sleep. But dream engineering studies designed to curate microdreams with specific content can then look at how these dreams impact creativity. Again, this ability to control dream content allows scientists to test whether dreaming affects performance independent of the effects of sleep.

In an earlier chapter, we met targeted dream incubation and the Dormio device, which uses sensors on the hand to detect sleep onset and verbally prompts subjects with specific dream themes as they fall asleep. After sleeping for a minute or two, the user is awakened to give a dream report, before returning to sleep for another round of microdreaming. A group of scientists at MIT conducted a study using this method to incubate the theme 'tree' into repeated sleep onset dreams for some subjects, while others fell asleep without the target theme (or a control group simply stayed awake).[7] After a period of forty-five minutes, both groups completed several tests of creative performance, related to the 'tree' theme. All of the subjects who slept and had repeated bouts of stage 1 sleep had better creative performance than subjects who stayed awake, for instance, telling more creative stories around the word 'tree', or finding more creative uses for a tree (such as a source of income for lumberjacks, a coffin, or a musical instrument). Subjects with tree-related microdreams outperformed those who did not dream of the target theme; in fact, creativity was directly correlated with the number of 'tree' references in dreams. So dreaming conferred a creative boost, beyond the benefits of stage 1 sleep alone (and simply thinking about trees while awake did not yield the same boost to creativity).

This confirms the intuition of many artists and scientists throughout history who have long touted the use of microdreams as a tool for creativity. It's becoming clear, both anecdotally and experimentally, that dream incubation, sleep onset microdreams, and both non-lucid and lucid dreams can all be used to seek insight or creative solutions to personal, logical, or artistic problems. And unlike the waking state, where we tend to systemically search through commonsense solutions, sleep offers more atypical answers, drawing together old ideas in new and unusual ways.

While Dormio offers an electronic approach, the basic technique used by Dalí (and Thomas Edison, too) for capturing hypnagogic insights requires simply dozing off with a heavy object in hand, and when muscle tone lessens and the object drops, you will awaken to recollect eureka moments from your dreaming mind. Even more simple is the 'upright napping procedure'[8] developed by dream scientist Tore Nielsen to entice a hypnagogic siesta at any time:

Step 1: When drowsy, sit upright in a chair, close your eyes and await a nap. Observe all thoughts and imagery while waiting.

Step 2: When a head nod or other muscle jerk triggers awakening, recall and record the details of any preceding imagery immediately.

Step 3: Repeat as many times as desired.

Try to focus on – or incubate – a specific problem while awaiting sleep, and immediately afterward record any observations with text or drawings. A momentary descent into hypnagogia might be just what you need for a creative solution to emerge. So beyond 'sleeping on it,' we should all set out to 'dream on it', too, in the pursuit of creativity.

Avatar: dreams and physical learning

In a multisensory 'dream hotel room,'[9] museum visitors in Switzerland slept in a six-legged bed that literally rocked them to sleep, under a sculpture of a mushroom (a 'fly agaric' mushroom) and a montage of moving red lights that induced the illusion of motion – a flying 'fly agaric' mushroom zooming overhead. As the flashing lights trickled into closed eyes, acoustic stimulation repeated the words 'flying with flying fly agarics' to incubate dreams of flying. Several museum visitors reported dreams of floating, falling, soaring and flying – both during their nights at the museum and over the following days – images of riding a roller coaster, swimming in water and coasting on a flying bird. While the whole experience was designed with the artistic in mind, it nicely demonstrates the embodied and multisensory nature of dreaming, too. And this brings us to another resource we can tap into in dreaming: learning to move our bodies within dream worlds.

As we established already, sleep is important for all sorts of learning, improving performance on everything from simple visuospatial tasks or vocabulary learning to playing instruments, downhill skiing, or other more real-world skills. Most sleep scientists think the brain is actually replaying recent memories to strengthen them while we sleep, and animal studies provide support for this idea: when rats learn to navigate a maze while awake, precise neural patterns appear in the hippocampus, encoding specific memories for places. Later when these rats are sleeping, the same neural patterns reappear, as if the brain is replaying its trajectory through the maze, learning how to navigate while asleep.

In humans, too, it seems that spatial memories are replayed and strengthened during sleep, exercising how and where to move in the world. And scientists can reinforce this replay process with targeted reactivation. As you now know, when sounds or smells are associated with memories prior to sleep, re-presenting these same cues during sleep strengthens the associated memories, thanks to the functions and neurophysiology of sleep. As an example, when an odour is paired with learning the precise locations of several objects, re-exposure to the odour during sleep leads to improved memory for these locations later.

Do reactivated memories also appear in dreaming? Do we literally traverse mazes in our nightly dream capades, and can we learn from this doing in dreaming?

Earlier I described a number of studies that show how dreaming of a particular task is linked to improved performance on that same task after sleep. But most of these studies were correlational; that dreams reflect learning does not mean that dreams *cause* learning. Again we turn to dream engineering, to see how manipulating dream content (to dream about a specific learning task) impacts later performance.

Key to this agenda is inducing task-related dreams on demand. To this end, many dream labs are now designing in-house learning tasks that are more readily incorporated into dreams, based on some of the scaffolds of dreaming we covered earlier. This is in contrast to older laboratory studies that used basic cognitive tasks, things like word learning or picture memorising, which are both unlikely to arise in dreams and difficult to detect; often less than 10 per cent of dreams incorporated learning tasks in such studies. Now we know more about the memory sources of dreaming: that people most often dream of memories that are novel and social, embodied and emotional. In the lab they often dream *about* the lab, the settings and personnel and tasks. So designing learning tasks around these scaffolds makes it more likely that they will later find their way into dreams.

Based on this knowledge, scientists at Northwestern University devised their own dream-worthy tasks based on real-world interactions with experimenters and environments that were at once highly novel, social, embodied and emotional.[10] These tasks consisted of learning to play a harmonica duet with an experimenter using only nasal inhalations, or competing against an experimenter to blow soap bubbles onto a target using only nasal exhalations (talk about dream-worthy!). The use of nasal inhalations and exhalations was by design, to attempt to observe these breathing signals during sleep as objective evidence of task replay. Each of the tasks was linked to a musical cue, and after engaging in the tasks in the evening, subjects got into bed and did some dream incubation, mentally rehearsing their performance while the corresponding musical cues played. During the night, the cue for only one of the tasks was presented in REM sleep, and subjects woke up several times to report their dreams.

Overall, almost all of the subjects dreamed of the tasks, and the cues boosted incorporation of the intended task into dreams. This shows again that targeted dreaming is possible, with scientists here injecting harmonica – or bubble – themes into these subjects' dreams. But are these targeted dreams useful? Can they enhance learning, for example?

This was the question asked in a similar study, where a custom-built virtual reality flying task (another dream-worthy task) was paired with targeted reactivation, to test whether inducing flying dreams would enhance subjects' performance following sleep. In the VR environment, subjects used handheld controllers to fly through landscapes – traversing a circuit of green circles while avoiding red circles – associated with rewarding sound cues that were re-presented in a nap shortly thereafter. There was a fourfold increase in flying dreams from before the study to the nap in the lab (1.7 per cent to 7.1 per cent), and over 10 per cent of dreams referenced flying later that night. The majority of these dreams (~80 per cent) were related in some way to the VR experience or setup, such as '*gliding at ground level near a mountain, I go back up, then down in a series of coloured circles.*'

As expected, those subjects who dreamed more of the task performed better following sleep. And targeted reactivation also helped: subjects performed better if the VR sounds were re-presented during sleep. Together, those subjects who underwent targeted reactivation *and* dreamed of the task performed the best.

Generally, the link between dreams and learning is increasingly robust. And while there is still much to learn, targeted dreaming could have far-reaching applications for real-world learning, revising and refining embodied skills in the nightly arena of dreams.

Though we've discussed a lot of different tasks so far related to problem-solving or motor learning, one of the more clinical applications of dream engineering could be in physical rehabilitation. For this, we can start to look beyond the simple stimuli elaborated so far – acoustics, movement, light – to more embodied ways of inducing sleeping sensations.

Electrical muscle stimulation is one such method where subtle electric shocks induce muscle contractions in the body, mimicking twitches that naturally occur during REM sleep. Remember that muscle twitches in REM sleep are like physical probes, where one muscle moves at a time, then several in succession – shoulder, elbow, wrist, finger. Against the relative stillness of the sleeping body, these twitches could help to calibrate the body map in the brain, which is especially critical during motor development. To this end, infants spend around eight hours in REM sleep each day, twitching through the one-to-one mapping of the cortex and over six hundred skeletal muscles in the body, gradually learning to control and coordinate their movements. These refined maps require maintenance and repair over time as our bodies grow and change, and especially so after an injury or stroke.[11]

Electrical muscle stimulation is one way to repair these maps. In fact, stimulating muscle twitches is already used in waking physical rehabilitation to help people regain motor function and coordination following injury. An open question is whether this kind of stimulation could also enhance motor learning during sleep, to complement its waking rehabilitative effects and capitalise on the natural motor function of twitches in REM sleep. This is quite a new area of research; one group in Chicago is testing whether electrical muscle stimulation can speed up physical recovery from stroke, while scientists in Switzerland are exploring whether and how these simulated twitches drive motor imagery in dreams. In both cases, the goal is to mimic the natural muscle twitches of REM sleep to directly affect sleep and dreams, and ideally to enhance motor learning, too.

Electrical stimulation can also be applied to other systems in the body, such as the vestibular system, where it induces sensations of shifting balance and motion. Electrodes placed behind the ears deliver a small electrical current to the three fluid-filled sacs in the ear canal (these create an x-y-z axis of the head, conveying our sense of movement, body position and balance). A small current can induce rather dramatic sensations of falling or spinning, and it's thought that vestibular activity underlies our illusions of bodily motion in dreams. Indeed movement imagery is one of the most common features of dreaming aside from the usual visual content, and this is despite the fact that the physical body is lying horizontal and still. While scientists hope to make use of vestibular stimulation to direct sensations of movement and balance in dreams, it has so far been difficult to titrate such stimulation to the sleeping body, as the resulting sensations vary dramatically and often provoke awakenings. Attempts to fine-tune this manner of stimulation, though, are underway.

Besides applying physical sensations to the body, we can also interface directly with the body map in the brain. Remember that in REM sleep, brain activity during dreamed body movements resembles brain activity during waking body movements. Because of this, one avenue to modulate dreaming is to directly modify activity in the motor cortex. A method called 'transcranial direct current stimulation (tDCS)' applies an electric current to the scalp to inhibit brain activity, and when applied during REM sleep, this decreased the proportion of dreams with movement. To elaborate, ten subjects spent two nights in the laboratory – one night with tDCS stimulation and one night without – and dreams during stimulation featured less repetitive motor activities like writing or walking, though they had just as many passive movements like riding in a car. This confirms that the sensorimotor cortex is in part a controller for dreamed body movements, and inhibiting the brain thereby inhibits dreamed movement.

What about when the real body is inhibited in some way, such as in the case of paraplegia? And what are dreams like for people who are blind or deaf? Does sensorimotor loss persist in the imaginary world of dreaming?

I often get asked about what dreams are like for those with sensory or motor loss in waking life, and there are generally two sides to my answer: yes, some loss does persist in dreaming, but it is also possible for such individuals to have dreams where the body and senses function fully. For instance, while people with paraplegia report fewer dreams with full body movements than control subjects, more than a third of their dreams still involve some voluntary leg movements, regardless of whether paraplegia was congenital or acquired after a spinal cord injury later in life. These dreams mostly include walking but also feature running, riding a bike, swimming, dancing, skiing and playing basketball. In one study, paralysed war veterans dreamed of walking in close to 50 per cent of their dreams, even after having been paralysed for over fifty years. And people who have had amputations also continue to dream of themselves as physically unharmed years after amputation.

For those who are blind, dream reports can and do include descriptions of visual activity, even if the subjects have been blind since birth (although whether this is phenomenally akin to experiences of vision in sighted subjects is hard to determine). Rapid eye movements also occur, albeit less so than in sighted subjects, and these eye movements correlate with the amount of visual imagery reported. Even before birth, fMRI data collected in utero reveals foetal eye movements that correspond with activity in visual brain areas, prior to any experience of vision. Given that rapid eye movements occur and visual dreams are reported in subjects blind since birth, this could represent an inherent capacity for sensory imagery, even in the absence of comparable waking sensation. Subjects who are deaf, too, report hearing spoken language in their dreams (though less so than hearing subjects), and those with hearing loss generally have more sensorially vivid dreams, with more smells and tastes relative to subjects with normal hearing.

What would be the purpose of such innate embodied dreaming? And what is the neural substrate of vivid dreams in those with sensorimotor loss in waking life?

In patients with spinal cord injury, dreamed movements

could be remnants of personal memories of pre-injury walking or sports activities. But what about the dreams of those paralysed since birth? As we know, the motor cortex is active when we imagine motor movements, and for paraplegics, this activation is higher when they attempt to move their limbs, suggesting a full motor program is available even if it cannot be fully executed. These motor programs can be learned simply through observing movements in others, creating sensory and motor representations in the brain that can be expressed in the dream body, regardless of the condition of the waking physical body.

In another famous example, author Helen Keller, who became deafblind before the age of two, wrote in her autobiography that before learning language, her dreams were 'devoid of sound, of thought or emotion of any kind, except fear, and only came in the form of sensations.' She remembers a recurring nightmare of '[running] into a still, dark room ... I felt something fall heavily without any noise, causing the floor to shake up and down violently; and each time I woke up with a jump.'[12] Curiously, after learning language in the form of tactile signs, Keller reports, 'As I learned more and more about the objects around me, this strange dream ceased to haunt me ... and I began to dream of objects outside myself.' Her imaginal capacity changed as she began to know more about the world around her, and as she learned about objects conceptually, she was then able to dream of them. In this case, perceptual representations learned through language gave shape to her dreams, even for perceptions never experienced before, only imagined.

There are other cases, too, where intact sensorimotor programs seem to be accessible in imagination, even if inaccessible in the external world. For instance, one man with brain lesions in the parietal cortex (important for motor control) was able to precisely control his finger movements only when he imagined these movements with his eyes closed, but when he tried to intentionally move his fingers, he was not able to. Somewhere in the brain his mind could access these movements, and his body could perform them, but there was a disconnect, a glitch that emerged during deliberate attempts of motor control. In a related example, while movement and speech difficulties are common to some waking disorders, like tremors in Parkinson's or poor vocal control in aphasia, these motor symptoms have been witnessed to temporarily resolve during sleepwalking and sleep talking. It's as if the brain maps and muscular skills are still available but become inaccessible or impeded when awake.

What these cases tell us is that it could be possible to tap into these intact sensorimotor programs during sleep, perhaps in an effort to repair them. Could methods like brain stimulation or lucid dreaming be used to curate more sensorimotor imagery in dreams? And if we can harness the brain's capacity for sensorimotor dreaming, can we use this to enhance physical training and rehabilitation in waking life?

The idea is not too far-fetched – in fact, waking motor imagery is already used in rehabilitation, reliably enhancing motor learning in patients recovering from injury. REM sleep, too, is associated with motor learning. Combining these concepts, in a study where subjects had their arm immobilised (placed in a sling) for twenty-four hours, motor imagery practice in waking life increased subsequent time spent in REM sleep, and benefited later motor control. In principle, then, mental imagery and REM sleep together support physical learning, even when physical training is not possible in waking life (as is often the case after stroke or injury). The next step in this research direction is to assess whether motor imagery in dreaming could further support physical rehabilitation.

We established already that lucid dreaming practice can benefit physical learning, such as on a darts-throwing task that requires fine-tuned limb coordination. We also saw that dreamed movements generate motor cortex activity similar to what we would observe during waking, like when lucid dreamers clench their left and right fist in a dream, it corresponds with activation in appropriate motor brain regions. So at least in the brain, moving the dream body is like moving our real body. Along these lines, one proof-of-concept study revealed that lucid dreamers can even control a brain-computer interface, using motor imagery within a dream to move a virtual object on a computer. Besides the cool factor of controlling a computer from your dreams, this research is relevant to the growing field of neuroprosthetics, where patients learn to control prosthetic limbs using only their imagination. In principle, when patients imagine moving their limb, it generates EEG signals in the motor cortex that are decoded by a computer interface and relayed to control the prosthetic movement. This is a type of brain-computer interface, where the brain communicates via computer to an external device.

In order to learn how to use neuroprosthetics, patients need hours and hours of practice in generating precise motor signals using mental imagery. Patients can also train in virtual environments – controlling the movements of a VR avatar with their mind. This helps them to refine their mental motor commands and to better control the real prosthetic later. Patients with spinal cord injury have even learned to control whole exoskeletons through this kind of virtual training. Whether patients could further exercise their mental muscles in the virtual arena of the dream world has not yet been studied, although dreaming is arguably more accessible and more embodied than VR, an innate and evolutionarily refined training ground available to pretty much everyone from the comfort of home.

Can we exercise our virtual dream bodies to regain motor control after stroke or injury or even to learn how to use a prosthetic limb? These might seem like futuristic notions for now, and though science has not yet attempted to bridge these gaps, I expect dream engineers will soon be tapping into and training sensorimotor programs in dreamland.

Eternal Sunshine: social dreamers

Moving from the physical body to the social and cultural world, let's explore: how far does dreaming extend into the social networks of our waking lives?

To start to answer this question we can visit sleep researcher Katja Valli and the Consciousness Research Group in Turku, Finland. The group is well known for their work on simulation theories of dreaming – that dreaming recreates worlds so that we can practise real-world skills. Chief among Valli and her group's focus is the social simulation theory, which describes how dreams emulate social reality to forge and fortify our social skills.[13] The social simulation theory, and the earlier threat simulation theory, both ascribe an evolutionary function to dreaming. According to these theories, dream content is biased, that is, it selectively represents certain features of our waking environments in order to support our survival as a species. Dreams portray social scenarios so that we learn to cooperate with others; dreams portray threats so we can learn to escape or fight back. Over time, these dream simulations strengthened our success as a species, affording a nightly playground to bolster our social and survival skills in waking life.

On my first day visiting the research centre in Turku a few years ago, Valli pointed to a movie poster on the wall, an image of a lone man on an island cast in shadows. The faux film poster was an homage to the lab's work on the remote island of Seili in the Turku archipelago. The island carried a dark history, housing first a leper hospital in the seventeenth century and after that a mental institution where patients were treated as inmates and confined to the island. Since the asylum's closure in 1962, the island has become a nearly deserted locale, except for a research facility where Valli and her team conducted their experiment: a study of how social seclusion impacts dreams. For the study, subjects were shipped to Seili and isolated in single rooms for four nights. They were provided with a notepad to be used for slipping notes under the door to personnel in case of emergency, but otherwise they were cut off from the social world.

Isolation is a condition that feels counter to everything

we know and expect as social animals. To many, it might sound like their worst nightmare, but the experiment was not as dire as it sounds. Despite the lack of social interaction, many subjects enjoyed the experience. They could spend their days as they liked, playing instruments, reading books and taking photographs of the island, with meals delivered three times a day. They also kept dream diaries throughout their stay and wore EEG monitors during sleep. What these measurements revealed was that, despite the isolation of each day, their dreams continued to be filled with social interactions. In other words, their dreams did not simply reflect the isolation of waking life but instead continued to be social, as if by design.[14] In theory at least, these social dreams could help us maintain and motivate social bonds in waking life, even when deprived of such activity in the prior days.

In this and other studies, people in seclusion (for instance, those in prison) dream more often of friends and family than those who are not socially isolated, suggesting that dreams nurture *close* attachments when these are lacking in waking life. Indeed, the social landscape of dreams is often woven of friends, family, and romantic partners, though we do also dream of strangers and unfamiliars when our waking relations are active and secure. In part the thinking is that, when we are not isolated and we do have the opportunity to nurture close relationships during the day, our dreams start to offer up more strangers and unfamiliar characters, pushing the function of our social brains to try out all kinds of interactions, even those uncommon to waking life. These scenarios develop our social competence as we sleep.

But social dreams are not always positive. In fact, if we think back to typical dream themes, relationship conflict is one of the most frequent bad dream themes that people report. These dreams can actually hinder social connection: dreams of jealousy and infidelity lead to more relationship conflict and reduced intimacy the next day. In a study where subjects kept daily logs of both their dreaming and waking interactions, it was specifically the night's dream content that predicted the next day's interactions (in other words, bad dreams *preceded* the waking conflict, and not the other way around). It's easy to imagine how these dreams can be difficult to brush off come morning and that lingering feelings of distress or distrust get in the way of intimacy the next day.

On the other hand, pleasant dreams of romantic partners lead to feeling more love and closeness towards them the next day. And dreams can also provide a sense of closeness to friends and family members who are absent in waking life. In fact, one poignant period of social *dis*connection where dreams can offer a form of solace is during the process of bereavement, when it is common to dream of lost loved ones. Perhaps surprisingly, grief dreams are more often positive than negative and sometimes take the form of what's called 'visitation dreams', dreams of seeing the deceased as healthy or happy, or acting in a way that is comforting, such as: *'My husband came and sat down by my bed and said "I've been to the end of time and back and you know what? I still love you."'*[15] These visitation dreams can bring a sense of closure and peace to someone in the process of bereavement, along with a real feeling of connection to the deceased.

While over 90 per cent of people who have had a romantic partner or pet die experience positive visitation dreams, it's also the case that for some, grief dreams immediately following loss can be distressing and nightmarish, with feelings of sadness, despair, or guilt. These dreams seem to accompany the trauma of loss and are similar to post-traumatic nightmares that potentially re-enact the memory of a death. In prenatal loss, for example, negative themes of seeing a baby dying or suffering or recalling a miscarriage are as common as positive themes of seeing a baby as healthy and happy. Between 25 per cent and 45 per cent of people report negative grief dreams, especially after traumatic loss, and the appearance of deceased family figures in a way that's upsetting – seeing someone as dead, dying, or ill – understandably causes distress and anxiety on awakening.

In general, a transition from negative to positive grief dreams seems to reflect a healthy process of bereavement. In this light, grief dreams can be informative, at least, if they are negative and can be especially helpful when they are positive. Nevertheless, an important consideration is that the meaning and impact of grief dreams, whether negative or positive, depends strongly on one's cultural background and beliefs.

For some cultures and religions, dreams of the deceased are seen as a window into a true spiritual realm of the dead rather than an indication of an individual's processing of grief. In a study of Cambodian refugees who survived the Pol Pot genocide, even positive visitation dreams were deeply upsetting because they revealed that the deceased had not yet been reborn. Cambodians believe that especially after a violent death, the spirit remains in a liminal state as a dream visitor or a ghost, and potentially a malicious one at that. Even after thirty-two years, over half of the refugees surveyed had frequent grief dreams (at least one in the previous month), and these dreams were associated with prolonged symptoms of grief, PTSD and anxiety. While some of the dreams seemed positive and nostalgic, featuring memories from a time when the deceased was happy and healthy, other dreams were more dire, with the deceased appearing unwell or replaying the trauma of their death. Both types of dreams led to sadness and even crying on awakening, reminding the dreamer of their loss and that their deceased relative was still not reborn. The most feared dreams were those where the deceased was calling for the dreamer to be with them; these were interpreted as potentially dangerous, putting the dreamer's soul at risk of leaving their body, resulting in possible illness or death.[16]

In this sociocultural context, visitation dreams do not provide solace for the dreamer, and the appropriate response in the dreamers' views was to perform rituals on awakening, such as leaving food on an altar for the deceased. An awareness of sociocultural contexts thus both precedes and informs the practical impact of such dreams on social well-being.

Other religious contexts also view the dream state as an access point to a dimension shared by the dead, as in the Tibetan Buddhist belief in the 'bardo' state – an intermediate state that we enter after death and in sleep and dreaming, when consciousness is less connected to the physical body. A practice called 'Tibetan dream yoga', which is similar to lucid dreaming, offers the potential for visitation dreams with old masters or spiritual gurus and is seen as a sign of progress on the spiritual path. Outside of this religious context, lucid dreamers also report using their dreams to visit with lost loved ones or even to elicit encounters with spiritual guides or the divine. It is possible that these kinds of lucid dreams could be therapeutic for the dreamer, offering a means of actively healing from loss and finding a sense of closure or peace.

In this way, lucid dreaming, like Tibetan dream yoga, can take the form of a spiritual practice. In Tibetan dream yoga, practitioners train in diverse skills for controlling the dream state at will – changing the size or shape of objects in the dream, making figures appear or disappear on command. Dream yogis then apply this mental dexterity to better control their mental contents in waking life – their emotions, habits and behaviours – treating all of these mental contents as if a dream, an illusion. The goal is to fully realise the illusory nature of reality and to maintain a high level of lucidity across all states of consciousness – waking, dreaming and sleeping – which will eventually aid in attaining awareness in the state after death. Experts of dream yoga claim to preserve awareness even in dreamless sleep, in the absence of any mental content. These experiences of dreamless sleep are reported to be profound and include reports of being immersed in white light and feeling a sense of oneness or nothingness.[17]

In lucid dreaming communities, experiences of lucid dreamless sleep are also reported and are more colloquially known as 'the void', often described as a black nothingness that the sleeper inhabits in the space between dreams, after the disintegration of one dream and before the next appears. Some experience the void as a minimal dream, with some sense of being present but in an empty space, while others lose all sense of self and describe the experience as pure awareness. While such topics are still poorly understood scientifically, it is clear that for some people, lucid dreaming, and dreams more generally, offer a portal to meaningful spiritual and social encounters – to visit with lost loved ones or spiritual teachers, or to attain connection to something greater than oneself.

All in all, our dreams are clearly informed by the cultural and spiritual milieu of our waking lives and are deeply interwoven with our nature as social beings.

With all this in mind, we can start to think about how to manage and potentially enrich the social dynamics of our dream worlds, perhaps as a means of healing our social well-being.

Dream sharing offers one way to build real-life social connections through dreaming and has proven helpful, broadly, for those facing terminal illness, undergoing divorce or bereavement, or suffering loneliness. To start with, patients facing terminal illness often suffer from feelings of isolation and can find it difficult to communicate with loved ones about the progress of their disease and fears around death and dying. Through sharing dreams, patients find it easier to broach topics like death and dying, through the metaphoric language of dreaming. In several studies, patients with cancer or AIDS found that dream sharing lessened feelings of isolation and helped them come to terms with the process of dying, feeling more prepared for death. Although dream sharing in this context can be emotionally intense, patients felt more optimistic and relaxed after discussions.

Studies suggest dream sharing can be beneficial when done with a health care provider, a caregiver, or in group settings, from cancer groups to grief counselling, among others. In group settings, dream sharing creates a sense of connection between group members and enhances feelings of intimacy and trust. This form of support can be especially valuable to those suffering social disconnection. For example, recently divorced women reported boosts to self-esteem and better self-understanding after participating in dream-sharing groups. Many of these women had frequent bad dreams around themes of being hurt, ridiculed, or ashamed; their social self-image suffered in the aftermath of divorce. While these dreams are distressing, they also provide salient material to work with, and sharing these dreams can lead to greater compassion for oneself and potentially for other group members, too.

To explore these effects experimentally, during my first research position at Swansea University in Wales, I spent some time studying the benefits of dream sharing for both personal insight and empathy – measurable facets of how well we understand ourselves and others. We asked participants to complete one-hour sessions of dreamwork following the Ullman method (so-named because it was developed by researcher Montague Ulmann in 1996), which invites one person to share their dream with either a partner or a group.[18] The group members then take turns asking questions and reflecting on the dream's content, finding links with the dreamer's waking life and uncovering some of the feelings and personal meanings of the dream.

Our findings were simple: discussing dreams led to gains in personal insight, unveiling information about real waking life issues. Subjects reported that they learned more about their life through discussing dreams in ways they would not have thought of on their own, and they came to realisations about how experiences from the past influence their present behaviour. These kinds of insights were more frequent after discussing dreams than after discussing waking life events, suggesting that the medium of dreams is especially useful for gaining insight into one's life. Discussing dreams with a partner also led to a boost in empathy towards the dream sharer, both immediately after a session and over the course of a couple weeks. While any personal discussions might have these benefits, Mark Blagrove – the study lead and my supervisor at the time – describes dream sharing as a particularly poignant means of self-disclosure: through the fictional and story-like facade of dreams we more readily disclose sensitive or personal topics that we might otherwise try to hide or avoid. This dream-supported self-disclosure helps us to better understand ourselves and others.

Altogether, it seems that dream sharing does facilitate the discussion of taboo or difficult topics in a way that feels safe and manageable, and this fosters deeper connections and empathy between people. More broadly, both dreaming and dream sharing can have practical benefits on our social well-being, acting as a support over the course of bereavement or when facing terminal illness, divorce, or even transient periods of isolation and loneliness.

Sleep On It: Dream Skills

If you are curious to try dream sharing on your own, the steps of the Ullman method can be followed quite simply either with a group or a partner:

Step 1: Read aloud a written account of a recent dream (in the present tense), allowing the group/partner to ask questions to clarify the contents and sequence of the dream.

Step 2: The group/partner then offers suggestions on how they would feel if this dream were their own and how this dream might reflect or be related to their own waking life. (For this step, group members start by saying, 'If this were my dream ...' before offering their reflections.)

Step 3: The dreamer then describes their own recent waking life, including experiences or personal concerns, and suggests how the dream might be connected to such concerns. All group members offer reflections at this stage, too, and at the end someone reads the dream back to the dreamer, to finish with a holistic experience and understanding of the dream.

Beyond simple dream discussions, some scientists, myself included, are now exploring the potential of 'dreamwork engineering',[19] where VR is used to help dreamers visually and spatially re-experience their dreams. In an exhibition in Austria, dreamers whispered their dreams into a microphone and dream objects were projected into virtual space in real time. This system is designed to help dreamers interact with their dream material in a manner more visceral and immersive than drawing or writing or speaking a dream. When paired with dream discussions, the goal is to help users more fully experience their dream and glean personal and social benefits from this process. The VR system can even tailor the experience based on the dream's social-emotional content, where soundscapes are added into the scene based on the tone of social interactions (friendly or aggressive, for example), and visual hues are mapped into the scene according to emotion (confusion is projected as an overlay of white fog, happiness a pale blue).

Overall, experiences of dream sharing, dreamwork and now dreamwork engineering can inspire insight and empathy and harvest social connection for those struggling with loneliness. Dreams both reflect and influence our social well-being – they can help us to process grief or even to face death – though the meaning and impact of such dreams vary widely across cultures.

All in all, dreaming offers a relatively untapped resource for multiple facets of well-being, from social health to physical learning and creativity, too.

Dreams can bestow a sense of connection and offer meaningful encounters, as seen in visitation dreams and spiritual dreams. Sharing dreams with others can foster empathy and build a sense of community, helpful for those feeling isolated, such as patients in hospice care, or bereavement, or more simply as a balm for loneliness in our post-pandemic society. Dreaming in these contexts at once pulls us inward, into the many lives of our sleeping mind, while also bridging outward, helping us relate to others through our shared experiences of dreaming.

Dreaming is also a fertile resource for artistic communities, useful for creators and inventors of all kinds. Dream incubation – reflecting on something just prior to sleep with the intention to dream about it – can reveal solutions to personal or creative problems. And science has now proven what artists like Salvador Dalí and inventors like Thomas Edison knew, that the ephemeral dreams from the sleep onset state are a reliable source of creativity, and they're simple to access, too. Modern approaches to harvest insight from dreams include technologies for inducing and capturing sleep onset dreams, weaving ideas into and out of dreamlands. On the other side of sleep, virtual reality can spruce up our interactions with dreaming in waking life, and dream discussions can reveal 'aha!' insights waiting just beneath a dream's content.

All of these uses of dreaming and dreamwork depend in part on the close ties between dreaming and memory – that dreaming is made of and remaking memory each night. Underneath our experience of sleeping life is a malleable and plastic memory system that evolves over the course of the night. Scientists are now attempting to mould dreams to support learning and rehabilitation, engineering embodied and movement-filled dreams for skill rehearsal. This area of research is ripe for discovery, and I expect that dream engineers of the future will design dreams to enhance physical learning, reinforce spatial memories, modify waking behaviours, and more.

I find it remarkable the progress dream science has made in just these past few decades, and still the momentum of the last few years leaves me feeling that the real discoveries have only just begun.

Conclusion

When we fall asleep, with the quieting of the external world, our imagination expands to fill our entire perceptual field and we journey through fully immersive and self-relevant realms.

Dreaming arises from reverberations through a web of memories, woven networks of experience, knowledge, emotions and sensations. In dreaming, we see ourselves: what we care about, who we interact with, where we are planning to go. We rebuild worlds based on waking experiences. These dreamscapes are set up in consistent ways, across cultures, across sleep studies, shaped by evolutionary pressures, and coloured by individual meaning.

Our dreams are also richly embodied; we feel our dream bodies as immersed in a sensory scene, filled with convincing characters and stories. Dreaming simulates reality, and our perspective as a first-person actor in this virtual world is so compelling that even when we know it is a dream it seems only more vibrant, engaging and endlessly surprising.

Our nightmares, like our dreams, are learned; we relive fearful memories and play out ill-fated scripts in the dream world. Most of us are not in the practice of examining nightmares; we prefer to avoid thinking about them, to avoid going back to sleep for fear of their recurrence. But by now it should be clear that the inner world of dreams and nightmares is malleable, suggestible to the influence of waking intention, lucid dreaming and even technologies for dream engineering. Treatments like imagery rescripting and lucid dreaming help those with nightmares to gain more mastery over how they feel in and respond to dreams, and to learn new stories to dream each night.

In 2022, as president of the International Association for the Study of Dreams, I had the honour of presenting a Lifetime Achievement Award to Stephen LaBerge while at a conference in Ashland, Oregon. In personal conversation, LaBerge described a turning point in his own path to facing nightmares through lucid dreaming. He realised quite simply that his feelings were key. 'You can't fool your dream,' he offered, because the dream is always wise to your true feelings, and as long as you are afraid, the nightmare will remain. As lucid dream scientist Clare Johnson would say, 'I am the dream and the dream is me';[1] we are simultaneously both producing and experiencing a dream, so the dream is reflecting back to us our feelings. Because of this conundrum, the trick to facing nightmares is to first find a way to feel calm, positive, open or curious so that when you turn to face the nightmare, it will become easily transformed, mirroring your positive feelings back to you.

When nightmares begin to improve, patients start to feel better both during sleep and when they wake up in the morning. Over time, these changes to our dreaming lives can start to reorganise the memories that are accessed within dreaming. The sleeping brain updates our memories each night, revising the knowledge we hold of ourselves and our place in the world. When nightmares are rescripted, the underlying memories re-form, too, altering how we relate to the past and deal with the present. In resolving nightmares, we can rewrite our autobiography and open the filter through which we perceive ourselves and the world.

Our experience of the waking world, too, is like a dream; or as Johnson says, *waking life is like a slower denser dream*.[2] Even though there is a real world out there that exists independent from our minds, it is also malleable and responsive to our actions and intentions, and, moreover, our personal experience of that world remains internal.

On a fundamental level, everything we experience – from our perception of everyday objects to our assumptions about other minds – is created in part by our own mind. We take two-dimensional sensory signals – forms of light, colour, sound, pressure – and construct a three-dimensional model of the world inside. As we walk through the external world, we remain immersed in an internal space, simulating, moment by moment, the very experience we appear to be having. Each of these moments is shaped by feeling, making the clouds more menacing on a moody day or more buoyant in a moment of laughter.

This is our dreaming mind. It permeates and co-creates our waking experience, weaving our feelings through every perception. And as we drift into sleep, sifting through a montage of memories, these are the moments that catch us – a snide comment at work, the cryptic look of a cashier, the enchanting colours of autumn leaves. These are the matter of dreaming.

Conclusion

Our sleeping and waking minds are not so different as we tend to think. It is in part the basis of dream engineering, that our sleeping bodies, too, function similarly to waking. Our senses continue to work at night – sound is transduced in the cochlea, light seeps into the retina, and so on. Of course, we encase ourselves in a relative sensory deprivation chamber – quiet, dark, still – to protect the fragile state of sleep (whereas in waking there is near-constant and abundant sensory input). But some sensory input still reaches the dreaming mind, and dream engineers use this stream of perception, however limited, to influence the content of dreams. The hue of a recent film, an audio clip of laughter, the scent of lavender – these subtle cues can act as reminders, nudging the dream in one direction or another, priming the mind to reinstate a familiar memory or to add a dose of calm or delight to a scene.

While the goal of working with dreams is to have more agency in how we move through the mental contents of sleep, lessons from dreamwork also translate to waking life. Of course, in the waking world, we do not have complete control over our environment (nor do we in dreams), but we *can* learn to manage our attention and our feelings, to bring more agency and intention into waking life – in short, to live more lucidly. Because our mind is co-creating our experience of the world, a lucid mindset can also illuminate waking life.

But are there limits to dream engineering?

Not long ago I was invited to the Toronto International Film Festival to speak after the screening of a new sci-fi film, *Daniela Forever*, about an experimental pill that induces lucid dreams as a potential cure for prolonged grief and depression. The main character, Nicolas, enters a clinical trial and starts to dream of his lover, Daniela, who died not long before. His mood and outlook brighten immediately. Over the course of the film, though, Nicolas becomes addicted to his dream world, confusing it with reality, mistaking his 'dream Daniela' for the real thing. This is the progression of many a dystopian dream sci-fi, where in tinkering with the dream world we end up losing parts of ourselves or losing sight of what is real. Like the erasure of true love's memory in *Eternal Sunshine of the Spotless Mind*, or the manipulation of one's inner thoughts in *Inception*, there is a blurring of lines between dreaming and reality. These fictional themes beg the question: are there risks to engineering dreams?

Back in the sleep lab, subjects sometimes ask if I can tell what they are dreaming about, if I can read their thoughts through the carefully placed electrodes on their scalp. For now, I assure them that I have no idea what they are thinking or dreaming of. But the quest for 'dream decoders' is already underway. In 2013, scientists were able to partly predict what a subject had been dreaming of using fMRI imaging. For the study, three subjects were awakened two hundred times (!) after brief sleep onsets in an MRI machine, and each time they were asked to report their dreams. Experimenters then identified any objects that appeared in these dreams and collected pictures of these objects, like houses or dogs or cars. The subjects then returned to the MRI machine, which measured their brain activity as they observed these pictures, while awake, of houses and dogs and cars. From the data gathered during this waking task, an algorithm was developed that could

Conclusion

guess which object a subject had been viewing based on their brain activity. When the same algorithm was applied to the original sleeping brain recordings, it was able to guess what the subject had been dreaming of.[3]

That's a fairly convoluted and cumbersome way to guesstimate what someone is dreaming about, but with recent advances in AI and neuroimaging, dream decoders will surely improve. On the one hand, this could be incredibly useful in dream science: to be able to study dreams without needing subjects to recall or report them, and even to collect measures of sleep experience all across the night without the hindrance of awakenings. On the other hand, this raises questions about the ethics of dream recording and whether subjects can truly give consent to the broadcasting of their subconscious mind.

Developments in dream recording are similar to advances in waking 'mind reading' – neuroimaging methods designed to deduce what someone is thinking about. Brain-computer interfaces use this technology to read subjects' motor signals, as we saw earlier, or even to decode language in patients who have lost the ability to speak. These are incredibly useful applications of mind reading. But potential misuses can also be concerning, such as collecting one's inner thoughts without their consent or the sharing of this data by companies.

In a similar vein, what if companies could incept their branding into dreams? In a recent controversy, advertisers attempted to market products in dreams using targeted dream incubation. In response, one legal paper argues that dream advertising is in breach of existing regulations, falling under the auspices of deceptive or subliminal advertising practices. The author argues that protections put in place around deceptive advertising in waking life should clearly translate to sleeping.[4] In general, there is a whole field of neurotechnology ethics that is proactively working to protect future subjects, and the public, from potential risks of mind reading or mind-influencing technologies – and any such policies would likely also apply to dreaming.

From a more positive angle, the translation of dream engineering techniques seems relevant to many different clinical contexts. One of my colleagues, Catherine Duclos, is studying whether brain stimulation methods used to augment dream lucidity could similarly raise conscious awareness in coma patients, helping them to reawaken from this unconscious state. Similar work has looked to dream science to better understand how to bring patients out of anesthesia, and new studies are even revealing the therapeutic potential of anesthesia dreams, where patients have reported overcoming traumas in transformative dreams that resemble nightmare rescripting in PTSD patients. In another potential crossover of knowledge: dream engineering could prove informative for psychedelic therapies, where clinicians rely on set and setting to guide a patient's journey. Dream engineering techniques are being refined to better guide the spontaneous world building of the mind and could similarly support the shaping of positive psychedelic experiences in therapeutic contexts.

Drawing someone out of a coma, tapping into anesthesia dreams, and directing psychedelic trips: these are all altered states of consciousness where some understanding

Conclusion

of how to interface with the dreaming mind could prove useful. After all, sleep itself is an altered state of consciousness, and one which we navigate through dreaming.

On a more general level, a movement towards more awareness and acceptance of dreaming as a core and consequential part of human life could follow naturally from the increased attention to sleep health that we've seen over the past few decades. In a way this is a return to the import dreams have had historically, where techniques of dream sharing and dream incubation, and engaging and interfacing with dreams, have been practised by cultures for millennia. In today's world, dream play can again become a part of culture through new dream-recording and dream-sharing apps, online (and even AI-driven) dreamwork communities, and portable dream engineering devices. What's more, with these growing online databases of dreams, scientists are ever more able to understand population trends in dreaming and to observe how dreaming is linked to public health on a larger scale. In the future, perhaps talking about dreams will be as usual as asking someone how their day is going or how well you slept last night, and as essential to a clinician as asking about physical health.

If there's one thing I know, it's that people love to talk about their dreams.

I remember my first research project at the age of twelve, when I presented a poster to the class on dreams – I demonstrated how our eyes move during sleep and told of the ancient Egyptians who used dreams for prophecy. But what I remember most clearly is what happened right afterwards: almost everyone in the class raised their hand to ask me about their dream. And to this day, this is the inevitable occupational hazard of telling anyone, 'I study dreams.'

In short, we are intrigued by our dreams and often feel compelled to share them. Dreams teach us, shape our waking minds, and help us to bridge our inner lives to other people and the world around us. And while I believe dreaming has always been one of life's great mysteries, it's one that I've spent the majority of my waking and sleeping hours exploring. *Never have I felt that science has been so close to cracking the mystery as now.*[5]

Notes

Introduction
1 S. Freud (1900), *The Interpretation of Dreams*, in *The Standard Edition of the Complete Psychological Works of Sigmund Freud*, vols. 4 and 5 (Hogarth Press, 1953).

1. The Scaffolds of the Dream World
1 C. Picard-Deland, T. Nielsen, and M. Carr, 'Dreaming of the Sleep Lab', *PloS One* 16, no. 10 (2021): e0257738.
2 A. Revonsuo, J. Tuominen, and K. Valli, 'The Avatars in the Machine-Dreaming as a Simulation of Social Reality', in T. Metzinger and J. M. Windt (eds.), *Open MIND* (MIND Group, 2015), 32(T).
3 T. A. Nielsen, A. L. Zadra, V. Simard, S. Saucier, P. Stenstrom, C. Smith, and D. Kuiken, 'The Typical Dreams of Canadian University Students', *Dreaming* 13 (2003): 211–35.
4 E. Aserinsky and N. Kleitman, 'Regularly Occurring Periods of Eye Motility, and Concomitant Phenomena, During Sleep', *Science* 118, no. 3062 (1953): 273–74.
5 K. M. T. Hearne, 'Lucid Dreams: An Electro-Physiological and Psychological Study' (doctoral dissertation, University of Liverpool, 1978); S. LaBerge, 'Lucid Dreaming: A Study of Consciousness During Sleep' (doctoral dissertation, Stanford University, 1980).
6 M. S. Blumberg, A. M. Plumeau (2016). 'A new view of "dream enactment" in REM sleep behavior disorder', *Sleep medicine reviews*, 30, 34–42.
7 C. Picard-Deland, M. Pastor, E. Solomonova, T. Paquette, and T.

Nielsen, 'Flying Dreams Stimulated by an Immersive Virtual Reality Task', *Consciousness and Cognition* 83 (2020): 102958.
8 E. Solomonova, 'The Embodied Mind in Sleep and Dreaming: A Theoretical Framework and an Empirical Study of Sleep, Dreams and Memory in Meditators and Controls' (doctoral thesis, McGill University, 2017); T. A. Nielsen, D. L. McGregor, A. Zadra, D. Ilnicki, and L. Ouellet, 'Pain in Dreams', *Sleep* 16, no. 5 (1993): 490–98.
9 T. Nielsen, 'Microdream Neurophenomenology', *Neuroscience of Consciousness* 2017, no. 1 (2017): nix001.
10 For a review of these early researchers, see: G. Morgese and G. Pietro Lombardo, 'History of Dream Research: Categorizations and Empirical Findings', in G. Morgese, G. Pietro Lombardo, and H. Vande Kemps (eds.), *Histories of Dreams and Dreaming: An Interdisciplinary Perspective* (Palgrave Macmillan, 2019), 247–74.
11 DeSanctis describes the formula: Dream = fundamental state of the dreamer (past experiences, intelligence, old habits) + momentary state of the dreamer (aspirations, concerns, health) + immediate experiences provoked by external conditions (stimuli during sleep). S. De Sanctis, 'L'interpretazione dei sogni', *Rivista di Psicologia* 10, nos. 5–6 (1914): 358–75.
12 S. LaBerge, B. Baird, and P. G. Zimbardo, 'Smooth Tracking of Visual Targets Distinguishes Lucid REM Sleep Dreaming and Waking Perception from Imagination', *Nature Communications* 9, no. 1 (2018): 3298.
13 D. Erlacher, M. Schädlich, T. Stumbrys, and M. Schredl, 'Time for Actions in Lucid Dreams: Effects of Task Modality, Length, and Complexity', *Frontiers in Psychology* 4 (2014): 1013.
14 Nielsen, 'Microdream Neurophenomenology'.
15 E. Olunu, R. Kimo, E. O. Onigbinde, M. A. U. Akpanobong, I. E. Enang, M. Osanakpo, I. T. Monday, et al., 'Sleep Paralysis, a Medical Condition with a Diverse Cultural Interpretation', *International Journal of Applied and Basic Medical Research* 8, no. 3 (2018): 137–42.

2. The Dreaming Brain
1 T. A. Nielsen, 'A Review of Mentation in REM and NREM Sleep:

Notes

'Covert' REM Sleep as a Possible Reconciliation of Two Opposing Models', *Behavioral and Brain Sciences* 23, no. 6 (2000): 851–66.
2. Aserinsky and Kleitman, 'Regularly Occurring Periods of Eye Motility, and Concomitant Phenomena, During Sleep'.
3. W. Dement and N. Kleitman, 'Cyclic Variations in EEG During Sleep and Their Relation to Eye Movements, Body Motility, and Dreaming', *Electroencephalography and Clinical Neurophysiology* 9, no. 4 (1957): 673–90.
4. W. D. Foulkes, 'Dream Reports from Different Stages of Sleep', *The Journal of Abnormal and Social Psychology* 65, no. 1 (1962): 14.
5. C. Speth and J. Speth, 'A New Measure of Hallucinatory States and a Discussion of REM Sleep Dreaming as a Virtual Laboratory for the Rehearsal of Embodied Cognition', *Cognitive Science* 42, no. 1 (2018): 311–33.
6. K. C. Fox, S. Nijeboer, E. Solomonova, G. W. Domhoff, and K. Christoff, 'Dreaming as Mind Wandering: Evidence from Functional Neuroimaging and First-Person Content Reports', *Frontiers in Human Neuroscience* 7 (2013): 412.
7. R. Stickgold, A. Malia, D. Maguire, D. Roddenberry and M. O'Connor, 'Replaying the Game: Hypnagogic Images in Normals and Amnesics', *Science* (2000), 290(5490), 350–353.
8. E. J. Wamsley, R. Stickgold, 'Dreaming of a Learning Task Is Associated with Enhanced Memory Consolidation: Replication in an Overnight Sleep Study', *Journal of Sleep Research*, 28(1), (2019): e12749.
9. M. Carr, M. Wary A. Grewal, S. Stafford, R. Raider, and W. R. Pigeon, 'Dreaming of the Sleep Lab Is Associated with Improved Performance on a Sign Language Learning Task: A Pilot Study', *Dreaming* (2023).
10. M. Schädlich, D. Erlacher, and M. Schredl, 'Improvement of Darts Performance Following Lucid Dream Practice Depends on the Number of Distractions While Rehearsing Within the Dream – A Sleep Laboratory Pilot Study', *Journal of Sports Sciences*, 35(23) (2017): 2365–2372.
11. D. J. Cai, S. A. Mednick, E. M. Harrison, J. C. Kanady, and S. C. Mednick, 'REM, not Incubation, Improves Creativity by Priming Associative Networks', *Proceedings of the National Academy of Sciences*, 106(25) (2009): 10130–10134.

12 R. Stickgold, L. Scott, C. Rittenhouse, and J. A. Hobson, 'Sleep-Induced Changes in Associative Memory', *Journal of Cognitive Neuroscience*, 11(2) (1999): 182–193.
13 P. Stenstrom, K. Fox, E. Solomonova, and T. Nielsen, 'Mentation During Sleep Onset Theta Bursts in a Trained Participant: A Role for NREM Stage 1 Sleep in Memory Processing?', *International Journal of Dream Research* (2012).
14 C. Picard-Deland, K. Konkoly, R. Raider, K. A. Paller, T. Nielsen, W. R. Pigeon, and M. Carr, 'The Memory Sources of Dreams: Serial Awakenings Across Sleep Stages and Time of Night', *Sleep*, 46(4) (2023): zsac292.
15 J. E. Malinowski and C. L. Horton, 'Dreams Reflect Nocturnal Cognitive Processes: Early-Night Dreams Are More Continuous with Waking Life, and Late-Night Dreams Are More Emotional and Hyperassociative', *Consciousness and Cognition* 88 (2021): 103071.
16 E. Solomonova, P. Stenstrom, T. Paquette, and T. Nielsen, 'Different Temporal Patterns of Memory Incorporations into Dreams for Laboratory and Virtual Reality Experiences: Relation to Dreamed Locus of Control', *International Journal of Dream Research* 8, no. 1 (2015): 10–26.
17 R. Wassing, O. Lakbila-Kamal, J. R. Ramautar, D. Stoffers, F. Schalkwijk, and E. J. Van Someren, 'Restless REM Sleep Impedes Overnight Amygdala Adaptation', *Current Biology* 29, no. 14 (2019): 2351–58.
18 M. P. Walker, 'The Role of Sleep in Cognition and Emotion', *Annals of the New York Academy of Sciences* 1156, no. 1 (2009): 168–97.
19 T. J. Cunningham, C. R. Crowell, S. E. Alger, E. A. Kensinger, M. A. Villano, S. M. Mattingly, and J. D. Payne, 'Psychophysiological Arousal at Encoding Leads to Reduced Reactivity but Enhanced Emotional Memory Following Sleep', *Neurobiology of Learning and Memory* 114 (2014): 155–64.
20 R. D. Cartwright, 'Dreams That Work: The Relation of Dream Incorporation to Adaptation to Stressful Events', *Dreaming* 1, no. 1 (1991): 3.
21 R. Cartwright, *The Twenty-Four Hour Mind: The Role of Sleep and Dreaming in Our Emotional Lives* (Oxford University Press, 2010).

Notes

3. Why Dream at All?

1. A. Damasio, *The Feeling of What Happens: Body and Emotion in the Making of Consciousness* (Houghton Mifflin Harcourt, 1999).
2. Author Joan Didion in her essay 'The White Album'.
3. Dream scientists Mark Solms, Sidarta Ribeiro, Eugene Gendlin, Ernest Hartmann, Antonio Zadra and Robert Stickgold, Josie Malinowski, and Tore Nielsen, among others, have theorised on the role of affect in dreaming.
4. 'The River of Consciousness' is an essay by the writer and neurologist Oliver Sacks, named so after psychologist William James in 1890 likened conscious experience to the flow of a river or stream.
5. Hartmann best describes this process of contextualising images, where emotion gives rise to image metaphors, in his book *The Nature and Functions of Dreaming* (Oxford University Press, 2011).
6. This example is from Eugene Gendlin's book on focusing and the felt sense: Eugene Gendlin, *Experiencing and the Creation of Meaning: A Philosophical and Psychological Approach to the Subjective* (Northwestern University Press, 1997).
7. D. Kahn, E. Pace-Schott, and J. A. Hobson, 'Emotion and Cognition: Feeling and Character Identification in Dreaming', *Consciousness and Cognition* 11, no. 1 (2002): 34–50; D. Skrzypińska and M. Słodka, 'Who Are the People in My Dreams? Self-Awareness and Character Identification in Dream', *International Journal of Dream Research* 7, no. 1 (2014): 23–32.
8. L. Kvavilashvili and G. Mandler, 'Out of One's Mind: A Study of Involuntary Semantic Memories', *Cognitive Psychology* 48, no. 1 (2004): 47–94.
9. Further discussion on this role of feelings in consciousness can be found in: M. Solms, *The Hidden Spring: A Journey to the Source of Consciousness* (Profile Books, 2021).
10. A. Zadra and R. Stickgold, *When Brains Dream: Understanding the Science and Mystery of Our Dreaming Minds* (W. W. Norton & Company, 2021).
11. Several authors view dreaming as a passive experience, for instance, describing dreams as 'single-minded' (A. Rechtschaffen, 'The Single-Mindedness and Isolation of Dreams', in *The Mythomanias* [Psychology Press, 2013], pp. 203–18); or that

specifically in REM dreams, 'to think is to do', meaning we act without reflecting in these dreams (Speth and Speth, 'A New Measure of Hallucinatory States and a Discussion of REM Sleep Dreaming as a Virtual Laboratory for the Rehearsal of Embodied Cognition'). Others highlight the possibility that thought, self-reflection and metacognition are rather commonplace in dreams (T. L. Kahan and K. T. Sullivan, 'Assessing Metacognitive Skills in Waking and Sleep: A Psychometric Analysis of the Metacognitive, Affective, Cognitive Experience [MACE] Questionnaire', *Consciousness and Cognition* 21, no. 1 [2012]: 340–52).

12 Co-creative theory views dreams as 'indeterminate from the outset and unfolding in real time according to the dream ego's moment-to-moment responses to the emergent content'. From: G. S. Sparrow and M. Thurston, 'Viewing the Dream as Process: A Key to Effective Dreamwork', *International Journal of Dream Research* 15, no. 1 (2022): 64–72.

13 J. J. Gross, 'The emerging field of emotion regulation: An integrative review', *Review of General Psychology*, 2 (1998), 271–299.

14 Neuroscientist Mark Solms describes this process where feeling moves us always towards coherence and homeostasis, a 'life-forward' direction, in his book *The Hidden Spring* (W. W. Norton & Company, 2021).

15 J. Zhao, S. F. Schoch, K. Valli, and M. Dresler, 'Dream Function and Dream Amnesia: Dissolution of an Apparent Paradox', *Neuroscience and Biobehavioral Reviews* 167 (2024): 105951.

16 K. Bulkeley, 'Dreaming Is Play', *Psychoanalytic Psychology* 10, no. 4 (1993): 501.

17 A. Cleeremans and C. Tallon-Baudry, 'Consciousness Matters: Phenomenal Experience Has Functional Value', *Neuroscience of Consciousness* 2022, no. 1 (2022): niac007.

18 T. Nielsen, 'The Stress Acceleration Hypothesis of Nightmares', *Frontiers in Neurology* 8 (2017): 201.

19 L. P. Marquis, S. H. Julien, A. A. Baril, C. Blanchette-Carrière, T. Paquette, M. Carr, J. P. Souxy, et al., 'Nightmare Severity Is Inversely Related to Frontal Brain Activity During Waking State Picture Viewing', *Journal of Clinical Sleep Medicine* 15, no. 2 (2019): 253–64.

20. M. Carr, K. Saint-Onge, C. Blanchette-Carrière, T. Paquette, and T. Nielsen, 'Elevated Perseveration Errors on a Verbal Fluency Task in Frequent Nightmare Recallers: A Replication', *Journal of Sleep Research* 27, no. 3 (2018): e12644; P. Simor, P. Pajkossy, K. Horváth, and R. Bódizs, 'Impaired Executive Functions in Subjects with Frequent Nightmares as Reflected by Performance in Different Neuropsychological Tasks', *Brain and Cognition* 78, no. 3 (2012): 274–83.
21. E. Hartmann, *The Nature and Functions of Dreaming* (Oxford University Press, 2014).
22. T. Nielsen and R. Levin, 'Nightmares: A New Neurocognitive Model', *Sleep Medicine Reviews* 11, no. 4 (2007): 295–310.
23. E. Hoel, 'The Overfitted Brain: Dreams Evolved to Assist Generalization', *Patterns* 2, no. 5 (2021): 100244.
24. V. Spoormaker, 'A Cognitive Model of Recurrent Nightmares', *International Journal of Dream Research* 1, no. 1 (2008): 15–22.
25. J. E. Malinowski and C. L. Horton, 'Metaphor and Hyperassociativity: The Imagination Mechanisms Behind Emotion Assimilation in Sleep and Dreaming', *Frontiers in Psychology* 6 (2015): 1132.
26. A. Revonsuo, 'The Reinterpretation of Dreams: An Evolutionary Hypothesis of the Function of Dreaming', *Behavioral and Brain Sciences* 23, no. 6 (2000a): 877–901; A. Revonsuo, 'Did Ancestral Humans Dream for Their Lives?', *Behavioral and Brain Sciences* 23, no. 6 (2000b): 1063–82.

4. When Are Nightmares a Problem?

1. M. R. Nadorff, D. K. Nadorff, and A. Germain, 'Nightmares: Under-Reported, Undetected, and Therefore Untreated', *Journal of Clinical Sleep Medicine* 11, no. 7 (2015): 747–50.
2. A. Gieselmann, M. Ait Aoudia, M. Carr, A. Germain, R. Gorzka, B. Holzinger, B. Klein, et al., 'Aetiology and Treatment of Nightmare Disorder: State of the Art and Future Perspectives', *Journal of Sleep Research* 28, no. 4 (2019): e12820.
3. American Psychiatric Association, *Diagnostic and Statistical Manual of Mental Disorders: DSM-5* (American Psychiatric Association, 2013).
4. W. E. Kelly, 'The Nightmare Proneness Scale: A Proposed Measure

for the Tendency to Experience Nightmares', *Sleep and Hypnosis* 20, no. 2 (2018): 120–27 (online).

5   V. Spoormaker, 'A Cognitive Model of Recurrent Nightmares', *International Journal of Dream Research* 1, no. 1 (2008): 15–22.

6   W. A. Youngren, N. A. Hamilton, K. J. Preacher, and G. R. Baber, 'Testing the Nightmare Cognitive Arousal Processing Model', *Psychological Trauma: Theory, Research, Practice, and Policy* 16, no. 8 (2023): 1401–8.

7   E. Solomonova, C. Picard-Deland, I. L. Rapoport, M. H. Pennestri, M. Saad, T. Kendzerska, S. P. L. Veissiere, et al., 'Stuck in a Lockdown: Dreams, Bad Dreams, Nightmares, and Their Relationship to Stress, Depression and Anxiety During the COVID-19 Pandemic', *PLoS One* 16, no. 11 (2021): e0259040.

8   R. J. Ross, W. A. Ball, K. A. Sullivan, and S. N. Caroff, 'Sleep Disturbance as the Hallmark of Posttraumatic Stress Disorder', *The American Journal of Psychiatry* 146, no. 6 (1989): 697–707.

9   R. L. Campbell and A. Germain, 'Nightmares and Posttraumatic Stress Disorder (PTSD)', *Current Sleep Medicine Reports* 2 (2016): 74–80.

10   W. Owczarski, 'Dreaming "the Unspeakable"?: How the Auschwitz Concentration Camp Prisoners Experienced and Understood Their Dreams', *Anthropology of Consciousness* 31, no. 2 (2020): 128–52; W. Owczarski, 'Adaptive Nightmares of Holocaust Survivors: The Auschwitz Camp in the Former Inmates' Dreams', *Dreaming* 28, no. 4 (2018): 287.

11   C. E. Titus, K. J. Speed, P. M. Cartwright, C. W. Drapeau, Y. Heo, and M. R. Nadorff, 'What Role Do Nightmares Play in Suicide? A Brief Exploration', *Current Opinion in Psychology* 22 (2018): 59–62.

12   J. Y. Que, L. Shi, L. Yan, S. J. Chen, P. Wu, S. W. Sun, K. Yuan, et al., 'Nightmares Mediate the Association Between Traumatic Event Exposure and Suicidal Ideation in Frontline Medical Workers Exposed to COVID-19', *Journal of Affective Disorders* 304 (2022): 12–19.

13   J. Belsky and M. Pluess, 'Beyond Diathesis Stress: Differential Susceptibility to Environmental Influences', *Psychological Bulletin* 135, no. 6 (2009): 885.

14   M. Carr and T. Nielsen, 'A Novel Differential Susceptibility Framework for the Study of Nightmares: Evidence for Trait

Sensory Processing Sensitivity', *Clinical Psychology Review* 58 (2017): 86–96.
15 E. Hartmann, *The Nature and Functions of Dreaming* (Oxford University Press, 2014).
16 E. N. Aron and A. Aron, 'Sensory-Processing Sensitivity and Its Relation to Introversion and Emotionality', *Journal of Personality and Social Psychology* 73, no. 2 (1997): 345.
17 J. M. Williams, M. Carr, and M. Blagrove, 'Sensory Processing Sensitivity: Associations with the Detection of Real Degraded Stimuli, and Reporting of Illusory Stimuli and Paranormal Experiences', *Personality and Individual Differences* 177 (2021): 110807.
18 E. L. Garland, B. Fredrickson, A. M. Kring, D. P. Johnson, P. S. Meyer, and D. L. Penn, 'Upward Spirals of Positive Emotions Counter Downward Spirals of Negativity: Insights from the Broaden-and-Build Theory and Affective Neuroscience on the Treatment of Emotion Dysfunctions and Deficits in Psychopathology', *Clinical Psychology Review* 30, no. 7 (2010): 849–64.

5. Treating Nightmares and Going Lucid

1 J. L. Davis, J. L. Rhudy, K. E. Pruiksma, P. Byrd, A. E. Williams, K. M. McCabe, and E. J. Bartley, 'Physiological Predictors of Response to Exposure, Relaxation, and Rescripting Therapy for Chronic Nightmares in a Randomized Clinical Trial', *Journal of Clinical Sleep Medicine* 7, no. 6 (2011): 622–31.
2 A. E. Kunze, A. Arntz, N. Morina, M. Kindt, and J. Lancee, 'Efficacy of Imagery Rescripting and Imaginal Exposure for Nightmares: A Randomized Wait-List Controlled Trial', *Behaviour Research and Therapy* 97 (2017): 14–25.
3 C. Schmid, K. Hansen, T. Kröner-Borowik, and R. Steil, 'Imagery Rescripting and Imaginal Exposure in Nightmare Disorder Compared to Positive Imagery: A Randomized Controlled Trial', *Psychotherapy and Psychosomatics* 90, no. 5 (2021): 328–40.
4 R. M. Gray and R. F. Liotta, 'PTSD: Extinction, Reconsolidation, and the Visual-Kinesthetic Dissociation Protocol', *Traumatology* 18, no. 2 (2012): 3–16.
5 C. R. Brewin, J. D. Gregory, M. Lipton, and N. Burgess, 'Intrusive

Images in Psychological Disorders: Characteristics, Neural Mechanisms, and Treatment Implications', *Psychological Review* 117, no. 1 (2010): 210.
6. Davis, Rhudy, Pruiksma, et al., 'Physiological Predictors of Response to Exposure, Relaxation, and Rescripting Therapy for Chronic Nightmares in a Randomized Clinical Trial'.
7. E. T. Gendlin, *Let Your Body Interpret Your Dreams* (Chiron Publications, 1986); Leslie A. Ellis, 'Solving the Nightmare Mystery: The Autonomic Nervous System as Missing Link in the Aetiology and Treatment of Nightmares', *Dreaming* 33, no. 1 (2023): 45; L. A. Ellis, 'Stopping the Nightmare: An Analysis of Focusing Oriented Dream Imagery Therapy for Trauma Survivors with Repetitive Nightmares', *Dissertation Abstracts International: Section B: The Sciences and Engineering* 76, no. 5-B(E) (2015).
8. Ellis, 'Stopping the Nightmare'.
9. G. Yount, T. Stumbrys, K. Koos, D. Hamilton, and H. Wahbeh, 'Decreased Posttraumatic Stress Disorder Symptoms Following a Lucid Dream Healing Workshop', *Traumatology* 30, no. 45 (2024): 550–59.
10. B. Holzinger, G. Klösch, and B. Saletu, 'Studies with Lucid Dreaming as Add-on Therapy to Gestalt Therapy', *Acta Neurologica Scandinavica* 131, no. 6 (2015): 355–63.
11. Mallett, R., Carr, M., Freegard, M., Konkoly, K., Bradshaw, C., & Schredl, M. (2021), 'Exploring the range of reported dream lucidity', *Philosophy and the Mind Sciences*, 2, 1–23.
12. S. LaBerge and H. Rheingold, *Exploring the World of Lucid Dreaming* (Ballantine Books, 1990), 24.
13. K. Appel, S. Füllhase, S. Kern, A. Kleinschmidt, A. Laukemper, K. Lüth, L. Steinmetz, and L. Vogelsang, 'Inducing Signal-Verified Lucid Dreams in 40% of Untrained Novice Lucid Dreamers Within Two Nights in a Sleep Laboratory Setting', *Consciousness and Cognition* 83 (2020): 102960.
14. A. Lemyre, L. Légaré-Bergeron, R. B. Landry, D. Garon, and A. Vallières, 'High-Level Control in Lucid Dreams', *Imagination, Cognition and Personality* 40, no. 1 (2020): 20–42.
15. R. Mallett (2020), 'Partial memory reinstatement while (lucid) dreaming to change the dream environment', *Consciousness and cognition*, 83, 102974.

16 Ibid.

6. Engineering Dreams

1. M. Carr, K. Konkoly, R. Mallett, C. Edwards, K. Appel, and M. Blagrove, 'Combining Presleep Cognitive Training and REM-Sleep Stimulation in a Laboratory Morning Nap for Lucid Dream Induction', *Psychology of Consciousness: Theory, Research, and Practice* 10, no. 4 (2023): 413.
2. S. LaBerge and L. Levitan, 'Validity Established of DreamLight Cues for Eliciting Lucid Dreaming', *Dreaming* 5, no. 3 (1995): 159.
3. G. Kumar, A. Sasidharan, A. K. Nair, and B. M. Kutty, 'Efficacy of the Combination of Cognitive Training and Acoustic Stimulation in Eliciting Lucid Dreams During Undisturbed Sleep: A Pilot Study Using Polysomnography, Dream Reports and Questionnaires', *International Journal of Dream Research* 11, no. 2 (2018): 197–202.
4. G. X. L. Zhang, 'Senses Initiated Lucid Dream (SSILD) Official Tutorial', 2013, available at http://cosmiciron.blogspot.com.au/2013/01/senses-initiated-lucid-dream-ssild_16.html.
5. R. Saredi, G. W. Baylor, B. Meier, and I. Strauch, 'Current Concerns and REM-Dreams: A Laboratory Study of Dream Incubation', *Dreaming* 7, no. 3 (1997): 195.
6. A. H. Horowitz, T. J. Cunningham, P. Maes, and R. Stickgold, 'Dormio: A Targeted Dream Incubation Device', *Consciousness and Cognition* 83 (2020): 102938.
7. Charlie Morley also led the online lucid dreaming workshop for PTSD patients described in chapter 5; https://www.charliemorley.com/.
8. S. Schwartz, A. Clerget, and L. Perogamvros, 'Enhancing Imagery Rehearsal Therapy for Nightmares with Targeted Memory Reactivation', *Current Biology* 32, no. 22 (2022): 4808–16.
9. C. Picard-Deland and T. Nielsen, 'Targeted Memory Reactivation Has a Sleep Stage-Specific Delayed Effect on Dream Content', *Journal of Sleep Research* 31, no. 1 (2022): e13391.
10. R. Wassing, O. Lakbila-Kamal, J. R. Ramautar, D. Stoffers, F. Schalkwijk, and E. J. Van Someren, 'Restless REM Sleep Impedes Overnight Amygdala Adaptation', *Current Biology* 29, no. 14 (2019): 2351–58.

11 A. Arzi, Y. Holtzman, P. Samnon, N. Eshel, E. Harel, and N. Sobel, 'Olfactory Aversive Conditioning During Sleep Reduces Cigarette-Smoking Behaviour', *Journal of Neuroscience* 34, no. 46 (2014): 15382–93.
12 N. D. Davenport and J. K. Werner, 'A Randomized Sham-Controlled Clinical Trial of a Novel Wearable Intervention for Trauma-Related Nightmares in Military Veterans', *Journal of Clinical Sleep Medicine* 19, no. 2 (2023): 361–69.
13 E. Peters, X. Wang, D. Erlacher, and M. Dresler, 'Targeted Lucidity Reactivation Using Muscle and Vestibular Stimulation', presentation for ISRW online conference, 2024.
14 K. R. Konkoly, K. Appel, E. Chabani, A. Mangiaruga, J. Gott, R. Mallett, B. Caughran, et al., 'Real-Time Dialogue Between Experimenters and Dreamers During REM Sleep', *Current Biology* 31, no. 7 (2021): 1417–27.
15 K. Appel, 'Advanced Communication from Lucid Dreams to the Waking World', Institute of Sleep and Dream Technologies, https://sd20.org/whitepapers/sweyepe_vo.1.pdf.
16 R. Mallett, L. Sowin, R. Raider, K. R. Konkoly, and K. A. Paller, 'Benefits and Concerns of Seeking and Experiencing Lucid Dreams: Benefits Are Tied to Successful Induction and Dream Control', *Sleep Advances* 3, no. 1 (2022): zpac027.
17 L. Aviram and N. Soffer-Dudek, 'Lucid Dreaming: Intensity, but Not Frequency, Is Inversely Related to Psychopathology', *Frontiers in Psychology* 22, no. 9 (2018): 384.

7. Bad Dreams and Health

1 C. Pierpaoli-Parker, C. J. Bolstad, E. Szkody, A. W. Amara, M. R. Nadorff, and S. J. Thomas, 'The Impact of Imagery Rehearsal Therapy on Dream Enactment in a Patient with REM-Sleep Behaviour Disorder: A Case Study', *Dreaming* 31, no. 3 (2021): 195.
2 J. M. Mundt, K. E. Pruiksma, K. R. Konkoly, C. Casiello-Robbins, M. R. Nadorff, R. C. Franklin, S. Karanth, et al., 'Treating Narcolepsy-Related Nightmares with Cognitive Behavioural Therapy and Targeted Lucidity Reactivation: A Pilot Study', *Journal of Sleep Research* (2024): e14384.
3 I often refer clinicians to this book to learn how to incorporate dreamwork into practice: L. Ellis, *A clinician's guide to dream*

Notes

therapy: Implementing simple and effective dreamwork (Routledge, 2019).
4   B. Sheaves, J. Onwumere, N. Keen, and E. Kuipers, 'Treating Your Worst Nightmare: A Case-Series of Imagery Rehearsal Therapy for Nightmares in Individuals Experiencing Psychotic Symptoms', *The Cognitive Behaviour Therapist* 8 (2015): e27.
5   P. A. Geoffroy, R. Borand, M. A. Akkaoui, S. Yung, Y. Atoui, Y., E. Fontenoy, J. Mauruani, and M. Lejoyeux, 'Bad Dreams and Nightmares Preceding Suicidal Behaviors', *The Journal of Clinical Psychiatry* 84, no. 1 (2022): 44174.
6   M. Gomes, M. P. Pérez, I. Castro, P. Moreira, S. Ribeiro, N. B. Mota, and P. Morgado, 'Speech Graph Analysis in Obsessive-Compulsive Disorder: The Relevance of Dream Reports', *Journal of Psychiatric Research* 161 (2023): 358–63; N. B. Mota, R. Furtado, P. P. Maia, M. Copelli, and S. Ribeiro, 'Graph Analysis of Dream Reports Is Especially Informative About Psychosis', *Scientific Reports* 4, no. 1 (2014): 3691.
7   G. Gopalakrishna, O. Popoola, A. Campbell, and M. A. Nemetalla, 'Two Case Reports on Use of Prazosin for Drug Dreams'. *Journal of Addiction Medicine* 10, no. 2 (2016): 131–33.
8   C. L. Nosek, C. W. Kerr, J. Woodworth, S. T. Wright, P. C. Grant, S. M. Kuszczak, A. Banas, D. L. Luczkiewicz, and R. M. Depner, 'End-of-life Dreams and Visions: A Qualitative Perspective from Hospice Patients', *American Journal of Hospice and Palliative Medicine* 32, no. 3 (2015): 269–74.
9   G. Schelling, C. Stoll, M. Haller, et al., 'Health-related quality of life and posttraumatic stress disorder in survivors of the acute respiratory distress syndrome', *Critical Care Medicine*, 1998; 26: 651–9.
10 S. T. Wright, P. C. Grant, R. M. Depner, J. P. Donnelly, and C. W. Kerr, 'Meaning-centered Dream Work with Hospice Patients: A Pilot Study', *Palliative & Supportive Care* 13, no. 5 (2015): 1193–211.
11 T. A. Nielsen, D. L. McGregor, A. Zadra, D. Ilnicki, and L. Ouellet, 'Pain in Dreams', *Sleep* 16, no. 5 (1993): 490–98.
12 M. Zappaterra, L. Jim, and S. Pangarkar, 'Chronic Pain Resolution After a Lucid Dream: A Case for Neural Plasticity?', *Medical Hypotheses* 82, no. 3 (2014): 286–90.

13 M. S. Parmar and Alejandro F. Luque-Coqui, 'Killer Dreams', *Canadian Journal of Cardiology* 14, no. 11 (1998): 1389–91.

8. Sleep On It: Dream Skills

1. D. Barrett, *The Committee of Sleep: How Artists, Scientists, and Athletes Use Dreams for Creative Problem-Solving – and How You Can Too* (Crown House Publishing Limited, 2001).
2. K. R. Konkoly, 'Dialoguing with Dreamers in REM Sleep to Explore Why We Dream' (doctoral dissertation, Northwestern University, 2024).
3. L. Roklicer, 'Lucid Dreaming for Creative Writing: Interviews with 26 Writers', *International Journal of Dream Research* 16, no. 1 (2023): 52–69; C. R. Johnson, *Llewellyn's Complete Book of Lucid Dreaming: A Comprehensive Guide to Promote Creativity, Overcome Sleep Disturbances & Enhance Health and Wellness* (Llewellyn Worldwide, 2017).
4. Https://dave-green.co.uk/.
5. S. Dalí, *50 Secrets of Magic Craftsmanship* (Courier Corporation, 1992).
6. C. Lacaux, T. Andrillon, C. Bastoul, Y. Idir, A. Fonteix-Galet, I. Arnulf, and D. Oudiette, 'Sleep Onset is a Creative Sweet Spot', *Science Advances* 7, no. 50 (2021): eabj5866.
7. A. H. Horowitz, K. Esfahany, T. V. Gálvez, P. Maes, and R. Stickgold, 'Targeted Dream Incubation at Sleep Onset Increases Post-Sleep Creative Performance', *Scientific Reports* 13, no. 1 (2023): 7319.
8. T. A. Nielsen, 'A Self-Observational Study of Spontaneous Hypnagogic Imagery Using the Upright Napping Procedure', *Imagination, Cognition and Personality* 11, no. 4 (1992): 353–66.
9. C. Holler, D. Hildebrand Marques Lopes, N. Di Chio, P. Maes, M. Carr, C. Picard-Deland, and A. Haar Horowitz, 'Flying Dreams Stimulated by Targeted Movement and Sound: Art and Science in The Dream Hotel', *Dreaming*, https://psycnet.apa.org/fulltext/2026-10153-001.html.
10. Konkoly, 'Dialoguing with Dreamers in REM Sleep to Explore Why We Dream'.
11. A. Tiriac, G. Sokoloff, and M. S. Blumberg, 'Myoclonic Twitching and Sleep-Dependent Plasticity in the Developing Sensorimotor

Notes

System', *Current Sleep Medicine Reports* 1 (2015): 74–79; M. S. Blumberg and J. C. Dooley, 'Phantom Limbs, Neuroprosthetics, and the Developmental Origins of Embodiment', *Trends in Neurosciences* 40, no. 10 (2017): 603–12.

12 H. Keller, 1903, *The story of My life*.
13 A. Revonsuo, J. Tuominen, and K. Valli, 'The Avatars in the Machine: Dreaming as a Simulation of Social Reality', in T. Metzinger and J. M. Windt (eds.), *Open MIND* (MIND Group, 2015).
14 J. Tuominen, H. Olkoniemi, A. Revonsuo, and K. Valli, '"No Man Is an Island": Effects of Social Seclusion on Social Dream Content and REM Sleep', *British Journal of Psychology* 113, no. 1 (2022): 84–104.
15 J. Black, T. DeCicco, C. Seeley, A. Murkar, J. Black, and P. Fox, 'Dreams of the Deceased: Can Themes Be Reliably Coded?', *International Journal of Dream Research* 9, no. 2 (2016): 110–14.
16 D. E. Hinton, N. P. Field, A. Nickerson, R. A. Bryant, and N. Simon, 'Dreams of the Dead Among Cambodian Refugees: Frequency, Phenomenology, and Relationship to Complicated Grief and Posttraumatic Stress Disorder', *Death Studies* 37, no. 8 (2013): 750–67.
17 A. Holecek, *Dream Yoga: Illuminating Your Life through Lucid Dreaming and the Tibetan Yogas of Sleep* (Sounds True, 2016); A. Alcaraz-Sanchez (2023), 'Awareness in the void: A microphenomenological exploration of conscious dreamless sleep' *Phenomenology and the Cognitive Sciences*, 22(4), 867–905.
18 M. Blagrove, S. Hale, J. Lockheart, M. Carr, A. Jones, and K. Valli, 'Testing the Empathy Theory of Dreaming: The Relationships Between Dream Sharing and Trait and State Empathy', *Frontiers in Psychology* 10 (2019): 1351; M. Blagrove, C. Edwards, E. van Rijn, A. Reid, J. Malinowski, P. Bennett, M. Carr, et al., 'Insight from the Consideration of REM Dreams, Non-REM Dreams, and Daydreams', *Psychology of Consciousness: Theory, Research, and Practice* 6, no. 2 (2019): 138.
19 P. Liu, A. Kitson, C. Picard-Deland, M. Carr, S. Liu, R. Lc, and C. Zhu-Tian, 'Virtual Dream Reliving: Exploring Generative AI in Immersive Environment for Dream Re-experiencing', in *Extended*

Abstracts of the CHI Conference on Human Factors in Computing Systems (May 2024), 1–11.

Conclusion

1. Johnson, *Llewellyn's Complete Book of Lucid Dreaming*.
2. Johnson, *Llewellyn's Complete Book of Lucid Dreaming*.
3. T. Horikawa, M. Tamaki, Y. Miyawaki, and Y. Kamitani, 'Neural Decoding of Visual Imagery During Sleep', *Science* 340, no. 6132 (2013): 639–42.
4. D. Marlan, 'The Nightmare of Dream Advertising', *William and Mary Law Review* 65 (2023): 259.
5. Quote by my supervisor, Tore Nielsen, in a personal email communication the summer before I started my PhD at the University of Montreal in 2010.

Acknowledgements

I'll start at the beginning then jump to the end, before traversing the in-between. First and foremost, I thank my parents for endlessly supporting my path, protecting my growth, and becoming my closest friends; I am forever grateful to you both.

I'm thankful to the publishing teams at Holt and Profile: Izzy Everington, I deeply appreciated your thoughtful and thorough input each step along the way; and Serena Jones, your voice and confidence gave me the strength to overcome those final hurdles. Kirsty McLachlan, thank you for championing the proposal and guiding me through the world of publishing.

My path in dream science has been inspired by many people. Tore: I feel so fortunate to have found my way to the DNL all those years ago, and I'm so glad I get to build my own lab alongside you. I'm also grateful to the CÉAMS for supporting dream science, and Antonio Zadra and Nadia Gosselin for encouraging my career. To my core dream team: Claudia, Rachel, Karen, Remy – I thank my lucky stars to have you as colleagues, companions and conference comrades around the world. To my dear and near friends in Montréal: Gaëlle and Farah, Vickie, Cloé, Liza, thank you for helping me find a home here.

Spanning the globe are countless mentors who filled my life with scientific adventure; especially thanks to Mark Blagrove and Wilfred Pigeon for supporting my path to independence, and to Katja Valli, Martin Dresler and Ken Paller. The IASD first opened my portal to dream studies, pointing me towards graduate school, supporting my early work, and carving a place for me as a leader, a mentor and a committed member of the field. Special appreciation for my dream guides – Angel and Kelly – and many others, too.

Into the Dream Lab

I also want to thank those who have inspired me as a scientist and as an author: Stephen LaBerge, Deirdre Barrett, Robert Waggoner, Ernest Hartmann, Josie Malinowski, Rosalind Cartwright, Oliver Sacks, V.S. Ramachandran, Matthew Walker, Sidarta Ribeiro and Mark Solms. And those who have helped me find deeper meaning in my dreams: Leslie Ellis, Clare Johnson, Andrew Holecek, Deborah Lewin, Eugene Gendlin and Charlie Morley. To the original DxE team – Adam Haar, Judith Amores, Pattie Maes and a shout to Pedro Lopes, thanks for building something new with me.

A little nod to the venues and vistas where I wrote: Rue Rivard, Chemin Beausoleil, and assorted MTL cafés.

To return at last to my family, thanks to: Brendan and Erin; Dan, Lena, Lucy, Stanley and Eugene; Gary Kraus; Ann McCarthy; and too many Carrs to name, for believing in me and for always bringing joy and comfort to my life; HC and Lu, my inner voices of encouragement, and the constant warmth and laughter in my heart; and Anatoly, mon âme, we've deftly travelled through time to be here, and I'm so delighted to share another of life's wondrous destinations with you.

Index

A
(becoming absorbed in fantasies) 131
addiction 229, 234, 251, 256–9, 267, 268, 270, 310
advertisers, dream 311–12
agency 5, 97, 151, 183, 191, 192, 228, 309
alexithymia 130
alien abduction 45
American Academy of Sleep Medicine 61
American Sign Language 71
amygdala 61, 81–3, 107, 114, 144, 181, 211, 213
anchoring, dream 170, 175
anesthesia 312–13
anxiety
  dreams 105
  disorders 82, 166
  grief dreams and 297, 298
  lucid dream therapy and 173, 177, 179, 221, 247, 254
  nightmares and 82, 121, 123, 125, 127, 129, 132, 134, 137, 138, 139, 142, 150, 166, 173, 177, 179, 251, 254

aromatherapy 214. *See also* scent
arousal 107–11, 113–16, 127–30, 133, 141–4, 158, 176, 181, 215, 217, 254, 259
  attention and 110, 111, 112, 113
  bodily 120, 127, 128, 141, 142, 167, 223, 240, 248
  dampened 118–20
  hyper 127, 130, 142, 143, 164, 167–8
  presleep cognitive 129
artificial intelligence (AI) 310–11, 313
Aserinsky, Eugene 56
attention
  arousal and 110, 111, 112, 113
  daydreaming style and 131
  dream recall and 46–50, 170
  dream themes/content and 13, 97
  lucid dreaming and 182, 184, 185, 189, 199, 201, 202, 206, 227
  managing 309

333

mindfulness and self-reflection, dreams high in and 150
multisensory integration and 43–4
nightmares and 145, 146, 147, 159, 172, 184, 252
process model of emotion regulation and 99
waking thought and 64
auditory stimuli 30, 36–8, 41, 59, 146, 201–4, 209, 210, 215, 217, 218, 223, 264, 289
autoimmune disease 260, 266
awakening 4, 7, 26
   bad dreams and 103, 117, 233
   central image and 113
   dream recall and 47, 48, 59, 60
   drug dreams and 257
   false 203, 226–7
   grief dreams and 297, 298
   lucid dreaming and 187, 194, 197, 203, 214, 226–7
   nightmares and 29, 103, 107, 117, 125, 129, 142, 150–51, 153, 157, 158, 179, 181, 187, 215, 233, 252
   sleep stages and 75, 77–8, 83–4, 107, 125
   sleep terrors and 243
   sleepwalking and 242
   upright napping procedure and 281
   vestibular stimulation and 287

B
bad dreams 233–69
   COVID and 132, 139
   exposure and 160–62
   focusing-oriented dreamwork and 172, 173, 175
   frequency of 117
   health and 233–69
   lucid dreaming and 7, 179, 180, 184, 187
   mental health and 250–60
   nightmares and 103, 117, 125, 126
   NREM parasomnias and 240–45
   overnight therapy, as 1
   physical health and 260–67
   REM parasomnia and 235–40
   REM rebound and 268
   sleep health and 235–50
   social design of dreams and 23
   term 233–4
   threat simulation theory and 119, 120, 125
'bardo' state, Tibetan Buddhism belief in 298–9
bathroom/toilet, trying to find within dream 26, 34–5, 36, 96
belly breathing 168, 174, 175
bereavement/grief dreams 30, 115, 125, 263, 272, 296–9, 300, 302, 304, 310
binding 41–2, 46

Index

bipolar disorder 254, 255, 256, 268
Blagrove, Mark 302
blindness 16, 30, 249, 288, 289
blood flow 38, 81, 108
body
    bodily arousal 120, 127, 128, 141, 142, 167, 223, 240, 248
    'body map' (sensorimotor cortex) 29–30, 32–3, 286, 288
    circadian rhythm disorders and 249
    design of dreams and 26–38
    dream engineering and 217, 218, 309
    dreams as products of our bodies 22–36, 50–51, 267, 268, 306
    embodied, dreams as richly 61, 72, 92, 170, 173, 175, 193, 229, 240, 271, 272, 282, 284, 286, 289, 293, 305, 306
    lucid dreams and 188, 189, 191, 202, 292
    memories and 97–100
    moving bodies within dream worlds 282–5
    muscle paralysis 27, 34, 35, 61, 238, 278
    muscle twitches 14, 28, 30, 32–6, 53, 54, 103, 216, 219, 237, 286–7
    nightmare and 3, 103, 142, 145–6, 148–9, 161, 167, 168, 169, 170, 214
    physical health, dreams and *see* physical health
    REM sleep and 118–19, 223, 240, 286, 287, 288
    temperature of 215
borderline personality disorder (BPD) 251, 253–4
boundary thinness 145
brain 52–85
    amygdala 61, 81–3, 107, 114, 144, 181, 211, 213
    architecture of sleep/sleep stages and 32–3, 53–65
    binding and 41–2, 46
    body map in 29–30, 32–3, 286, 288
    brain-computer interface 292–3, 311
    cognitive function *see* cognitive function
    creativity and 273, 278, 279
    delta waves 54, 63
    dreaming not confined to 8–9, 27
    drug dreams and 258
    electrical brain stimulation 216–17, 291, 312
    electrodes to measure activity in 3, 14, 15, 36, 40, 63, 245–6, 287, 310
    emotion and 80–84
    engineering dreams and 193, 194, 195, 196, 204, 213, 214, 215, 216–17, 223, 229
    first-night effect and 36
    high dream recallers and 46–7
    hybrid state and 242

imaging *see* neuroimaging
learning and *see* learning
lucid dreaming and 181, 183
medial prefrontal cortex 46, 63–4
memory and see memory
mirror neuron system and *see* mirror neuron system
motor cortex 30, 59, 61, 288, 290, 292
multisensory integration and 43–6
muscle twitches influence 32–3, 36
night watch and 36
nightmare and 2, 3, 4–5, 7, 104, 106–11, 114, 118, 119, 120, 121, 144, 146, 149, 158
prefrontal cortex 41–2, 46, 63–4, 107, 183
RBD and 236–7
recalibration (sensorimotor) 33
reward systems 80
sensorimotor regions 29–30, 31, 33, 271, 288, 289–91, 293
sleep apnoea and 248
spindles 54
theta waves 53–4
ultradian cycle 55, 56
white matter density 46

C
Cambodian refugees 297–8
cardiovascular health 124, 141–3, 147
Cartwright, Rosalind 84

cataplexy 238
central image 113, 163, 171
chaos in dreams 172, 173
characters, dream 19–20, 25, 59, 306
  feelings and 98, 99
  inaccurate representation in dreams 93
  lucid dreaming and 163, 170, 182–3, 189, 191, 275–6
  memory and 65
  mirror neuron system and 31, 32
  partners and 112–13
  schizophrenia and 255–6
  sensory stimuli during sleep and 37, 38
  social landscape of dreams and 295
chases within dreams 23, 64, 119, 133, 168, 178, 188, 208
chronic pain 229, 260, 264–5, 268
cigarette smoking 38–9, 212
circadian rhythm 55–6, 62
  disorders 249
clearing a space 169–70
clonazepam 268–9
clothing, dreams of inappropriate 24
co-creative dreaming 96–7
cognitive function 42, 69, 92, 94, 172, 224, 237, 284
  cognitive appraisal 109–10
  cognitive avoidance 129
  cognitive control 109–12, 158, 182, 183, 192

Index

cognitive health 52
cognitive neuroscience 7, 8
  dream recall and 57
  learning and 72
  lucid dreams and 182, 183, 192, 219
  nightmares and 128–31, 144, 148, 150–51, 158, 167
coma 312
consciousness 60, 89, 92, 95, 100, 102, 186, 250, 298–9, 312–13
Consciousness Research Group 293
cortisol 142
COVID-19 pandemic 131–3, 139, 304
CPAP machine 248
creativity 10, 120, 131, 149, 153, 204, 269, 304, 305
  co-creative dreaming 96–7
  creative potential of dreaming 131, 149, 269, 270–82
  hyper-associativity and 75
  lucid dreaming and 191, 270, 271, 274–7, 280
  nightmares and 9, 152, 159, 163, 270
  problem solving, dreams providing creative solutions to 83–4, 85, 93, 117

D

Dalí, Salvador 277–9, 281, 303
Damasio, Antonio 89

Daniela Forever (film) 309–10
databases, online dream 313
daydreaming 63, 64, 65, 95, 96, 148, 152, 153, 266
  guilty-dysphoric daydreams 130–31
  positive constructive daydreaming 130
  suffering martyr daydream 130–31
  waking daydreaming style 130–31
daymares 131, 166–7
day residue 6, 77, 79, 80
De Koninck, Joseph 65
deafness 30, 104, 288, 289, 290
decision-making in dreams 182
declarative and nondeclarative memories 66–7, 69
decoders, dream 310–11
Dement, William 56
depression 84
  antidepressant medications 269
  clues to progression or recovery within dreams 267
  imagery rescripting and 166, 250
  insomnia and 260
  lucid dream therapy and 247, 310
  nightmares and 123, 125, 138, 139, 141, 142, 147, 247, 251, 253, 254, 256

design of dreams 13, 14–51, 118, 181, 194, 270, 305
  attention and 46–50
  body and 26–38
  goals, dreams reflecting and 20–22, 24
  how dreams are designed 14–26
  microdreams and dream imagery 42–4
  multisensory integration 43–6
  performative nature of dreams 20
  repetitive or circular, dreams as 21
  scaffolding of our everyday life, dreams as 13
  sensory incorporation 36–9
  skill rehearsal or achievement 20, 21–2, 50
  sleep lab and 14–17, 19, 20–21
  social nature of dreams and 19–20
  themes, 'typical' dream 23–6
  waking perception or imagination, dreaming as more akin to 39–42
diatheses 143–4, 147–8
diary, dream 47–8, 150, 185, 295
differential susceptibility 143, 147
dissociation 82, 130, 131, 228, 253
disrupted/disturbed sleep
  diagnosis of disorders and 254–5, 266–7
  inducing lucid dreams through intentional 185–7, 190, 197, 199, 216, 225–9
  insomnia and 246
  nightmares and 2, 4–7, 107–8, 118, 120–21, 141–3, 149, 152, 167, 181, 194, 229, 246, 266
  presleep cognitive arousal and 129
  REM parasomnia and 236–8
  scent and 213
  sleep studies and 36, 81
divorce 84, 131–2, 300, 301, 302
Dormio 204, 206, 280, 281
Dream and Nightmare Laboratory, Montreal 52
dream control 187–92, 206
dream-enacting behaviours 30–32, 268
Dream Engineering Lab, Montreal 16, 200–201
dream functions. *See individual function name*
dream-lag effect 79
dream-reality confusion 226, 239–40, 242
'dreaming', stigma attached to term 250
Dreaming-sleep or D-sleep 57
DreamLight 199

Index

dreamwork 10, 169, 170, 172, 173, 269, 270, 301, 303, 304, 305, 309, 313
drug dreams 234, 256–9
Duclos, Catherine 312
dying patients 262

E

eating disorders 166, 251, 259
Edison, Thomas 281, 305
EEGs (electroencephalograms) 63, 223, 242, 245, 292, 295
  EEG headband 206, 209, 222
  high-density EEG 63, 242, 245
electrical brain stimulation 216–17
electrical muscle stimulation 216, 286–7
emotion, dreaming and 80–85
  emotion dysregulation 158, 254
  emotional cascade 112
  emotional reactivity 108–9, 128
  epic dreams and 249
  feelings and 90–91, 98–9
  lucid dreaming and 181, 182–4, 188, 191, 277
  memory and regulation of 81–5, 89, 90, 98, 101
  nightmares and regulation of 2, 7, 30, 103–22, 124–30, 133–4, 141, 144–5, 147–52, 158, 160, 162, 165, 175, 181, 192
  overnight therapy, dreams as and 80–85
  process model, emotion regulation 99, 182
  scent and 213
  sleep to forget/sleep to remember model 81
  stages of sleep and 59–61, 62, 65, 74, 80, 81–5, 181, 211, 269
  targeted reactivation and 211
engineering dreams 8, 193–230, 310
  devices used at home 221–2
  Dormio device 204, 206, 280, 281
  Dream Engineering Lab, Montreal study 200–201
  dream engineering research network 221–2
  dream incubation 203–8, 212, 280, 281, 284, 305, 311, 313
  DreamLight 199
  electrical brain stimulation 216–17
  electrical muscle stimulation 216, 286–7
  entrainment 195, 217
  eye-movement messages 216, 218–20
  false awakenings 203, 226–7
  Fluid Interfaces Group and 221–2
  international workshop on dream engineering, first (2019) 222

limits to 309–10
lucid dreaming and 193, 196–203, 206, 210–29
Muse or Dreem headband 222–3
Night-Ware 214
nocebo effect, sleep health wearables and 224
pleasurable dreams, attempts to induce 209–11
presleep training period 201
recording dreams 217–18, 221–3
rescripting therapy and 206, 208–9, 211, 221
scent stimuli and 211, 212–17, 229
self-efficacy and 205–6
side effects of controlling dreams 225
side effects of lucid dreaming 225–6
sleep trackers, downside of 224
targeted dream incubation 203–8
targeted forgetting 211–12
targeted reactivation 195–7, 199, 203–204, 209–212, 228, 229
temperature, sleep and 215, 217
two-way dialogues with lucid dreamers 218
vestibular stimulation 216
vibration and 214–15, 217, 222, 223
virtual reality (VR) 210, 220–21
wake-induced lucid dreams 202–3
within-sleep stimulation alone, ways of directing dreams using 213–16
wearables 204, 214, 221–4
ZMax headband 222
entrainment 195, 217
epic dreaming 249
episodic memory 66, 73, 74, 113–14
erotic dreams 23–4, 39, 240
Eternal Sunshine of the Spotless Mind (film) 311
evolution 1, 3, 6, 55, 119–20, 168, 293, 294, 306
exam, dreams of being late for/missing an 21, 25, 184
exploding head syndrome 33–4
exploration, dream 22, 23, 24, 25, 50, 95, 101, 102, 120, 121, 149, 172, 279
exposure, relaxation and rescripting therapy 160–62, 164–5, 167, 168, 170, 209, 252, 265
external environment, awareness of the during dreaming 36, 128
eye-movement during dreaming
lucid dreaming and 28, 199, 216, 218–20

Index

measuring 14, 199
rapid eye movements *see* REM sleep
saccades 40
sending messages with 28, 216, 218–20
smooth-pursuit pattern 40
Stage 1 sleep and 53

F

falling, dreams of 23, 30, 33, 35, 36, 105, 152, 179, 202, 266, 282, 287
fear extinction 114, 120, 161–2
fear of sleep 136–7, 138, 143, 208
feelings 4, 9, 89–122
   dream content, giving rise to 9, 89–102, 147, 307, 308
   feeling of knowing 93–4
   intensity of 90, 96, 103, 106, 107, 109, 111, 113, 115, 116, 117
   natural dream process and 90–102
   nightmares and 102–22, 147–8, 158–9, 169–71
   valence of 89, 94, 103
fever dreams 215
fight, flight, or freeze 109, 111, 168
'finding the help' 169, 171–3
first-night effect 36
Fluid Interfaces Group, The 221–2
flying dreams 30–31, 35, 36, 39, 50, 102, 151, 188, 189, 204, 210, 282, 285

focusing-oriented dreamwork 169–70, 173
Foulkes, David 57, 58
Freud, Sigmund 6, 77
functional magnetic resonance imaging (fMRI) 30, 81, 211, 289, 310

G

goals, dreams reflecting 20–22, 24, 92, 139
Green, Dave 277
grounded, feeling of being 170, 206–7
guided meditation 206
guilty-dysphoric daydreams 130–31

H

habits, personal dream 6
Hartmann, Dr Ernest vii, 145
healing dreams 177, 191, 192, 206, 207, 265, 299, 300
hearing in dreams/auditory stimuli 30, 36, 37–8, 41, 59, 146, 175, 201–4, 209, 210, 215, 217, 218, 223, 264, 289
Hearne, Keith 28
heart rate 3, 14, 29, 54, 82, 91, 7, 119, 141, 142, 167, 204, 214
heartbeat-evoked potential 149
Hebbian process 74, 76, 113
high dream recallers 46
hippocampus 114, 213, 283
Holocaust survivors 132, 134–6

hospice care, dream sharing in 263
hotel room, dream 282
hybrid state 242, 278
hyper-associativity 73, 75, 76, 83, 85, 118, 271, 279
hypnagogia 281–2
hypnic jerks 33
hypnotherapy 243

I
imagery rehearsal therapy 160–76, 196, 209–10, 237, 239, 243, 254, 265
implicit learning 271–2
Inception (film) 272, 273, 310
incubation, dream 203–4, 205, 212, 280, 281, 284, 305, 311, 313
inefficacy, dreams of 133, 139
infantile amnesia period 104–5
infidelity 23, 91, 105–106, 272, 296
inner worlds 5, 8, 90, 99–100, 148, 152, 307
insight (dream) 76, 77, 149, 151, 181, 271, 272, 273–82, 301, 302, 304, 305
insomnia 16, 49, 123, 129, 138, 139, 229, 245–7, 250, 253, 260, 268, 270
Institute of Sleep and Dream Technologies, Germany 219
Institute of Sport Science, Switzerland 216

intensive care units (ICUs) 262–3
ICU delirium 263
intention setting 47, 71, 172, 180, 190, 193, 196, 200, 201, 228, 247, 305, 308, 309
intentional sleep disruption 185–6
International Association for the Study of Dreams 307
conference (2012) 233

J
Johnson, Clare 307, 308
journal, dream 47, 183–4, 185, 190, 239

K
karaoke study 81, 211
Keller, Helen 290
killer dreams 265–6
kinesthetic imagery 70
Kleitman, Nathaniel 56
Konkoly, Karen 197

L
LaBerge, Stephen 28, 186, 199, 307
lavender 213–14, 309
learning
consolidation of skills via dreaming 6, 7, 10, 21–2, 52, 65–72, 85, 100–101, 160, 196, 269, 271–2, 282–93, 294, 302, 304, 305, 306
implicit 271–2

Index

learned behaviour 6, 160, 306
  lucid dreaming and *see* lucid dreaming
  memory and *see* memory
  physical 269, 282–93, 304, 305
  sleep/dreaming and "sleep on it" effect 273
  state-dependent learning 92–3
  unconscious 66, 89, 272, 274
  unlearning 77, 115–16, 119
leg-cuff stimulation 37
life-forward direction 100
lucid dreaming 3, 10
  benefits of 151, 247, 270
  breathing in a dream, ability to control and 28–9
  chronic pain and 264, 265
  control, learning how to 187–92
  creative potential 276–8
  devices and 222–4
  dream incubation and 203–4
  engineering lucid, pleasant dreams 196–212
  eye-movement messages and 28, 216, 218–20
  flying dreams and 210
  induce, learning how to 183–7
  learning tasks and 71–2, 292
  interfacing with the dreamer 44, 213–21
  lucid dysphoria 228
  lucid nightmare 227
  negative consequence of practicing 225–8
  nightmares and 7–8, 9, 148, 149, 151, 160, 176–92, 193, 196, 197, 200, 203, 205–212, 213–17, 219–22, 227, 228–30, 238–40, 247, 307
  orgasm and 240
  perception relatively true to waking life 39–41
  senses initiated lucid dream technique 201–2
  sensorimotor brain activity in waking life and 30
  sensorimotor imagery and 291–2, 299
  sleep onset microdreams and 206–7
  sleep quality and 247, 270
  smooth pursuit and 40
  as spiritual practice 299–300
  stimuli types used to induce 215–20
  therapy 176–92
  time perception in 172
  two-way dialogues with lucid dreamers 218–20
  VR and 220–21
  wake-induced lucid dreams 202–3

M

Maes, Pattie 221–2
maladaptive beliefs 164

McCartney, Paul: 'Yesterday' 273
medial prefrontal cortex 46, 63–4
medication 258, 268–9
meditation 182, 185, 206
memory 8, 13, 14, 19, 39, 42, 50, 65–80
   consolidation 81, 82, 85, 98, 196
   day residues and 6, 77, 79, 80
   declarative and nondeclarative 66–7, 69
   dream-lag effect and 79
   dream recall and 46–7
   emotion and 80–82, 84–5, 90
   episodic 66, 73, 74, 113–14
   feelings and 92–101
   forgetting/erasing irrelevant memory traces 68, 81–3, 211–12
   Hebbian process and 74, 76, 113
   learning and 66–72, 77, 81–3, 92–3, 282–4
   long-term 39, 66, 80
   lucid dreaming and 186, 194, 195, 196, 204, 211–12, 213, 239
   malleability of 101, 305
   nightmare and 105, 106, 113–21, 130, 133–4, 136–41, 161, 165–6, 306–8
   priming and 75
   prospective 186
   reactivation 68, 81, 97
   semantic 66, 67, 74
   sensory features of dream imagery, adds context and form to 44–7, 50, 52
   sequential model 67–8
   sleep stages and 67–83, 211–12, 273
   sleep to forget/sleep to remember model 81
   sources of 77–80, 106, 284
   state-dependent memory 92
   types 66–7
Mendeleev, Dmitri 273
mental health 4, 9, 58, 192, 196, 234, 244
   addiction 256–60
   psychiatric disorders 250–56
   nightmares and 1–3, 122, 123, 126, 138–43, 150–52, 176, 230, 250–54, 256, 258–60
   See also individual disorder name
microdreams 42–4, 205, 206, 207, 208, 212, 277–81
mindfulness 150, 182, 185, 247
mind pops 94
mind reading 311, 312
mind wandering 4, 63–5, 131
minimal dreams 48, 49, 300
mirror neuron system 31–2
mirror-tracing 46
MIT Media Lab, workshop on dream engineering at (2019) 222

Index

mixed dream-reality perception 242–3
money, dreams of finding 24
Morley, Charlie 206
motor cortex 30, 59, 61, 288, 290, 292
motor imagery in dreaming 29–30, 67, 237–8, 265, 282–3, 286–93
motor neurons 29
multisensory integration 43–4
muscle paralysis 27, 34, 35, 61, 238, 278
muscle twitches 14, 28, 30, 32–6, 53, 54, 103, 216, 219, 237, 286–7
Muse or Dreem headband 222–3

N

Nadorff, Michael 138
narcolepsy 16, 237–40, 270
near-infrared spectroscopy 108–9
neuroimaging 30, 31, 46, 59, 81, 108, 211, 245–6, 289, 310, 311
neuroprosthetics 292–3
neuropsychiatric lupus 266
neuroticism 127, 143
Nielsen, Tore 42–3, 52, 281
nightmares
  alexithymia and 130
  arousal/hyperarousal and 107–11, 113–15, 116, 118, 120, 127–30, 133, 141–4, 158, 164, 167–8, 176, 181, 215, 217, 223, 240, 248, 254, 259
  avoid, attempts to 96, 104, 126, 136–7, 138, 157, 158, 162, 164, 167, 182, 208, 269, 307
  bad dreams and 103, 117, 125, 126
  blood flow to frontal brain regions and 108
  boundary thinness and 145
  cardiovascular health and 124, 141–3
  central images and 113, 163, 171
  chronic 1–2, 106, 107, 108, 120–22, 138, 141, 142–3, 152, 163–4
  cognitive appraisal/cognitive control and 109–12
  dampened arousal and 118–20
  daymares 131, 166–7
  deafness and 104
  defined 103
  diathesis-stress framework and 143
  disrupted sleep and 4–5, 7, 107–8, 142–3, 149, 152, 181, 237
  emotion dysregulation and 106–7, 254
  emotional cascade and 112
  emotional reactivity and 108–10, 127, 128
  fear of sleep and 136–8, 143, 208

feelings and 102–6, 109–13, 121, 127, 130, 145, 147–50, 158–9, 162, 169–72, 175, 182, 188
frequency 3, 106, 107, 117, 120, 125, 126–7, 129, 133, 134, 137, 140, 141, 148, 162, 164, 165, 176, 193, 209, 220, 239, 248, 252, 260, 262, 266
health consequences of 126, 138–43
idiopathic 105
infantile amnesia period and 104–5
insomnia and 123, 129, 138, 139, 245–7, 253
lucid dreams and 7–8, 9, 148, 149, 151, 160, 176–92, 193, 196, 197, 200, 203, 205–212, 213–17, 219–22, 227, 228–30, 238–40, 247, 307
mental health and 1–3, 122, 123, 126, 138–43, 150–52, 176, 230, 250–54, 256, 258–60
neuroticism and 127, 143
nightmare disorder 124–31, 138, 140, 143, 158
nightmare proneness 127
'Nightmares: Under-Reported, Undetected, and Therefore Untreated' 123
older adults and 126–7
parents' separation and 105–106
pathological 25, 120, 124, 125, 131, 134
perpetrator trauma and 137–8
personality traits and 124, 126–9, 145
post-traumatic stress disorder (PTSD) and 29, 124, 133–43, 163–8, 176–7, 205, 213–15, 221, 228–9, 248, 249, 250, 251, 268–9, 312
presleep cognitive arousal 129–30
psychiatric disorders and 140, 251–2
reminiscent stimuli and 129–30, 133, 134
repetitive themes 105, 115–16, 209
risk factors 2, 138, 141
sensory processing sensitivity and 128, 144
sleep disorders and 142, 235–6, 239, 240, 245, 249
sleep loss and 108, 121, 253
stress and 2, 7, 9, 29, 103, 105–19, 121, 125, 127, 129, 131–4, 140–44, 147–8, 158, 160, 161, 167–8, 173, 175, 176, 181–3, 214 *see also* post-traumatic stress disorder (PTSD)
thought-suppressors and 128–9
threat simulation theory and 119–20
trauma and *see* trauma

Index

treating 124, 139, 149–50, 157–92, 193
dream engineering and 197–212
embodiment exercise 170–71
exposure, relaxation and rescripting therapy 160–62, 164–5, 167, 168, 170, 209, 252, 265
focusing-oriented dreamwork and 169–70, 173
imagery rehearsal therapy and 160–76, 196, 209–10, 237, 239, 243, 252, 254, 265
lucid dreams, overcoming nightmares through 176–92
nighttime therapy steps 174–6
   Step 1. Establish a sense of safety and relaxation 174
   Step 2. Recall and write down or tell (or draw) the nightmare 174–5
   Step 3. Rescript the nightmare 175
   Step 4. Rehearse the new dream 175
psychiatric disorders and 252–4
PTSD and 140–41, 143
rescripting and *see* rescripting therapy
self-efficacy and 205–6
targeted dream incubation and 206–208
upside of 1, 9, 124, 143–53
waking lives and *see* waking lives
Night-Ware 214
night watch 36
nocebo effect 224
non-REM (NREM) sleep 54–8, 61–3, 65–8, 70, 72–3, 75, 80, 217, 240–46, 248
NREM parasomnias 240–45
noradrenaline 81–2, 119
Northwestern University 220, 238–9, 274–5, 284
novel creations, dreams as 50
nude, dreaming of being 24, 25

O

obsessive-compulsive disorder (OCD) 255
olfactometer 211, 222
olfactory system 213
orexin neurons 238
orgasm 240
overnight therapy, dreams as 1, 80–81, 83, 211
Owczarski, Wojciech 135
'ownership' upon entering dream world 38

P

paraplegia 288–9
pareidolia 146
parents' separation 105–106

Parkinson's disease 234, 236, 237, 267, 291
partner
  dreaming of 23, 91, 105–106, 112, 241, 261, 272, 296, 297
  dream sharing with 170, 302–3
pathological nightmares 25, 120, 124, 125, 131, 134
patterns in disturbed dreaming 254–5
perception (dream) vs. imagination 39–41
performance
  creative 274, 277–8, 280
  dream lab, stress of 20–22, 24–5, 111
  dreams as performative 20
  lucid dreaming and 39–40, 271
  nightmare sufferers and 144–5, 158
  task performance, dreaming correlated with improved 69–70, 72, 85, 274, 277, 280, 282, 283, 284, 285
perpetrator trauma 137–8
personality traits, nightmares and 124, 126–9, 145
phantom limb dreams 292, 293
phantom word illusion 146
phenomenal experience, similarity and importance of waking and dreaming 39–41

physical health 124, 139, 141–3, 234, 260–67, 313
  rehabilitation of 272, 286–93
Pigeon, Wilfred 260
placebo effect in dream work 214, 224
polysomnography (PSG) 52, 221, 222, 235, 245, 279
positive/pleasurable dreams 7, 229
  bad dreams and 117
  benefits of 149–52, 153
  inducing 208–11
  lucid dreaming and 151, 187, 191, 247, 227
  motherhood and 261
  nightmare sufferers and 147–50, 152, 162, 165, 174, 196
  positive constructive daydreaming 130–31
  PTSD and 177
  rehearsing 208
  rescripting and 162
  scents and 194
  themes 23–4, 35
post-traumatic stress disorder (PTSD)
  body temperature and 215
  ICU nightmares and 263
  insomnia and 260
  lucid dreaming therapy and 176–7, 228
  nightmare therapy and 163–8, 168, 176–7, 205, 213–15, 221, 228, 229

Index

nightmares and 29, 124, 133–43, 163–8, 176–7, 205, 213–15, 221, 228–9, 248–51, 268–9, 312
  scents and 213–14
  sleep apnoea and 248
  VR and 221
prazosin 258–9, 268–9
prefrontal cortex 41–2, 46, 63–4, 107, 183
pregnancy 234, 260–61, 267
prenatal loss 297
presence in the room, sensing a 24, 238
presleep 19, 71, 83–4, 195–6, 203, 223
  cognitive arousal 129–30, 168
  insomnia and 246
  training 199, 200–201, 213, 247, 254
priming 75, 309
process model, emotion regulation 99, 182
prodromal dreams 265–67
psychedelic therapies 312–13
psychosis 227–8, 229, 251–55, 268

R

rapid eye movements 18, 28, 34, 54, 119, 289. *See also* REM sleep
reality testing 185, 190, 200, 216, 220, 221, 227–8
rebound effect 83, 128, 162
recalibration (sensorimotor) 33
recording dreams 20, 28, 40, 46, 47, 217–18, 221–3, 311, 313
recurrent dreams 243, 244
  epic dreams and 249
  lucid dreams and 225–6
  nightmares and 105, 115–16, 209
  REM sleep behaviour disorder (RBD) and 34, 236, 237
  skill rehearsal and 21–2, 120, 125, 135, 137, 157, 163, 166
rehearsal, dreams and
  imagery rehearsal therapy 160–76, 196, 209–10, 237, 239, 243, 252, 254, 265
  skill rehearsal 20, 21–4, 26, 50, 70, 72, 101, 119–20, 133, 270, 272, 284, 292–6, 300–302, 304, 305
  social simulation theory 20, 23–4, 26, 50, 101, 120, 133, 272, 292–6, 300–302, 304
relaxation 167, 244, 247, 265, 278, 279
  exposure, relaxation and rescripting therapy 160–62, 164–5, 167, 168–70, 209, 243, 252, 265
REM (rapid eye movement) sleep 18
  apnoeas and 248
  brain-to-body map and 33
  creativity and 273, 279, 284
  defined/explained 54–5, 60–61

electrical brain stimulation and 288
electrical muscle stimulation and 286
emotion and 80–84
engineering dreams and 197–8, 199, 200–201, 209, 211, 216, 222, 223, 226
forgetting distress associated with a memory and 81
hyper-associative nature of 73, 75–6, 83, 85, 271, 279
lucid dreaming and 181, 182, 185–6, 190, 197, 199, 200–201
medication to suppress 268–9
memory and 67–8, 70, 72–7, 79–82, 113, 114, 115, 273
misperception and 246
morning amplitude 55, 56, 60, 62–3, 70, 124–5, 186
muscle paralysis and 27, 30, 32, 33, 35, 61, 238
muscle twitches and 32–3
nightmares and 107, 111, 113, 114, 115, 125, 127, 136, 141, 149, 161, 181, 268–9
noradrenaline and 81–2
phasic and tonic 54
physical arousal levels during 118–19
physical learning and 291–2
REM parasomnia 235–40
REM rebound 268
REM sleep behaviour disorder (RBD) 34, 38–9, 236–7, 268
vivid and elaborate dreams, sleep stage associated with most 56–8, 60–63, 64, 65
reminiscent stimuli 129–30, 133, 134
rescripting therapy
  chronic pain and 265
  dream engineering and 206, 208–9, 211, 221
  drug dreams and 258
  epic dreaming and 249
  exposure, relaxation and rescripting therapy 160–62, 164–5, 167, 168, 170, 209, 252, 265
  imagery rehearsal therapy and 160–76, 196, 209–10, 237, 239, 243, 254, 265
  lucid dreaming and 176–83
  nightmares and 160–69, 171–2, 175–8, 180, 181, 188, 191, 238–9, 253, 266, 307, 308, 312
  psychosis and 253
  shift work and 250
  targeted reactivation and 211
  VR and 221
reward systems 80
rheumatoid arthritis 266
Rice, Anne: *The Queen of the Damned* 273
rooms or passages in your home, discovering 24, 25

Index

S

saccades 40
safety (rescripting theme) 162, 164, 168–70, 172, 174, 192
salience 89, 91, 107, 109, 145, 152
scaffolding of our everyday life, dreams as 8, 13, 20, 23, 50, 99, 104, 270, 284
scanning hypothesis 28
scent stimuli 37, 90, 194, 195, 196, 211, 212–17, 229, 309
schizophrenia 251, 255–6
scuba diving 92–3
seclusion, people in 294, 295
Seili, Turku archipelago 294
self-consciousness 24, 81
self-efficacy 205–6
self-report questionnaires 141, 145, 147, 235, 247
semantic memory 66, 67
senses initiated lucid dreaming technique 201–203
sensorimotor brain regions 29–30, 31, 33, 271, 288, 289–91, 293
sensory incorporation 35–8
sensory processing sensitivity 128, 144
set and setting 312
sexsomnia 240
sharing, dream 178, 263, 300–304, 311, 313–14
shift work 137
  disorder 249–50
Siclari, Francesca 245

side effects
  controlling dreams 225
  lucid dreaming 225–6
  sleep medications 269
signs, dream 184–5, 187, 190
skills, dream 270–305
  bereavement/grief dreams and 272, 296–9, 300, 302, 304
  blindness and 288, 289
  bodies within dream worlds, learning to move 282–3
  brain-computer interfaces and 292–3
  Cambodian refugees and 297–8
  creativity and 270–82, 304, 305
  deafness and 288, 289, 290
  dream sharing and 300–304
  electrical muscle stimulation and 286–7
  hypnagogia and 281–2
  inducing task-related dreams on demand 283–4
  learning and 269, 271–2, 282–93, 294, 302, 304, 305, 306
  limits of 272
  lucid dreaming and 270, 271, 274–7, 281, 291, 292, 299–300
  neuroprosthetics and 292–3
  paraplegia and 287–8
  partner cheating or leaving, dreaming about 272

physical learning 282–93
physical rehabilitation 286–93
prenatal loss and 297
seclusion, people in 295
sensorimotor skills 271, 282–92
sharing, dream 300–304, 311, 313–14
sleep onset microdream and 277–81
slumber with a key technique 277–8
social skills 293–300
Stage 1 sleep and 277–81
targeted dreaming 275, 276, 284–6, 311–12
threat simulation theory and 294
Tibetan Buddhist beliefs and 298–9
upright napping procedure 281–2
virtual reality and 285, 293, 303–4
visitation dreams and 296–9, 304
skill rehearsal 20–26, 50, 70, 72, 101, 119–20, 133, 270, 272, 284, 292–6, 300–302, 304, 305
sleep apnoea 16, 142, 247–9
sleep architecture 53–65. *See also* stages of sleep
sleep disorders 16, 49, 135–6, 142, 239, 240, 245, 247, 249, 250. *See also individual disorder name*

sleep drunkenness 240
sleep health 124, 224, 234, 235–50, 313
  circadian rhythm disorders 249–50
  epic dreaming 249
  insomnia *see* insomnia
  nightmares and 249
  NREM parasomnias 240–45
  REM parasomnias 235–40
  sleep apnoea 16, 142, 247–9
sleep hygiene 175–6, 235, 245
sleep lab 7–8, 14–17, 79, 197, 206, 212, 238–9
  architecture of sleep and 53, 54
  basic principles of 14–17
  dreaming of 19–23, 79
  first-night effect and 17–20, 36
  similarities in subjects' dreams 17–19
  See also individual lab name
sleep loss 108, 121, 225, 228, 253
sleep masks 199, 215, 221–2
sleep medicine 49, 61, 84, 224, 234, 235–69, 270
  mental health 250–60
  physical health 260–67
  renewed and growing interest in 235–6, 267–9
  sleep health 235–50
sleep misperception 49, 245, 246
sleep onset 4, 9, 33, 43, 54, 69, 70, 204–8, 215, 222, 246, 305, 310

Index

sleep onset microdream 277–81
sleep paralysis 26–7, 45, 226–8, 238, 249
sleep-related eating 240, 243
sleep scores 224
sleep talking 240, 241, 291
sleep terrors 240, 243–5, 249
sleep trackers 222, 224
sleepwalking 16, 240, 241–4, 291
slow wave sleep 54, 56, 63–4, 68, 80, 214, 215, 217
'slumber with a key' technique 277–8
smooth pursuit 40
social dreams
   nightmares and 145–7, 150
   social anxiety disorder 166
   social design of dreams 19–20, 23–4, 26, 62
   social simulation theory 20, 23–4, 26, 50, 101, 120, 133, 272, 292–6, 300–302, 304
   stages of sleep and 62
   VR and 304
Solomonova, Elizaveta 224
spatial navigation 30, 46, 67, 69, 71, 73, 282, 283, 303, 305
SPECT scanner (single-photon emission computed tomography scanner) 108
speech graph 255
spinal cord injury 288–9, 293
Spoormaker, Victor 128–9
stages of sleep 3, 15, 18, 53–65, 67–8, 72–3, 78, 222, 223

non-REM (NREM) *see* non-REM (NREM) sleep
REM *see* REM sleep
Stage 1 53–5, 59, 75, 77, 277–81
Stage 2 53–4, 59–61, 64, 70, 279
Stage 3 54, 55, 60, 63
Stevenson, Robert Louis 273
Stickgold, Robert 69
stress
   bad dreams and 1, 267
   nightmares and 2, 7, 9, 29, 103, 105–19, 121, 125, 127, 129, 131–4, 140–44, 147–8, 158, 160, 161, 167–8, 173, 175, 176, 181–3, 214
   post-traumatic stress disorder (PTSD) *see* post-traumatic stress disorder (PTSD)
substantia nigra 236–7
suffering martyr daydream 130–31
suicide 2, 123, 138–9, 142–3, 251–4, 256, 260, 270
Swansea University 301

T
targeted dreaming 275, 276, 284–6
   targeted dream incubation 203–8, 280, 311–12
   targeted forgetting 211–12
   targeted reactivation 195–7, 199, 203–204, 209–212, 228, 229, 239, 274, 283, 285

task-related dreams 70–72
  failing to complete tasks 23
  inducing on demand 283–4
teeth-falling dreams 26, 35, 36
temperature, sleep and 215, 217
temporal perception in dreams 40–41
Tetris 69–70
thalamus 213, 238
themes, dream
  baby-themed dreams 260–61, 297
  bad dreams and 119, 132, 234, 296, 301
  Covid-19 pandemic and 133, 139
  eating disorders and 259
  lab dreams and 19–23
  narcolepsy and 238
  nightmares and 104–5, 106, 160, 162–3, 164, 168, 172, 174, 175, 184, 252
  positive 150, 158, 203, 209–10, 280, 285
  recurring 6, 7, 24–5, 105, 106, 160, 163, 172, 174, 184, 187, 190
  rescripting *see* rescripting
  social dreams and 296, 297, 301
  'typical' 23–6, 34–6, 41, 50, 60, 96, 103, 115, 119, 133, 148, 158, 167, 180, 190, 245, 259, 296

theta rhythms 53–4
thirst in dreams 31, 94, 95
thought-suppressors 128–9
threat simulation theory 119–20, 294
Tibetan Buddhism 298–9
training, sleep and nightmare 235, 251
trait-state model 143
transcranial direct current stimulation (tDCS) 288
trauma
  nightmares and 9, 19, 29, 104–6, 113–16, 122, 124, 129–43, 142, 152, 153, 163–8, 176–7, 205, 211–15, 221, 228–9, 248–51, 262–3, 268–9, 312
  post-traumatic stress disorder (PTSD) *see* post-traumatic stress disorder (PTSD)
tree theme, study to incubate into repeated sleep onset dreams 280–81

U
Ullman method 301–2
ultradian cycle 55, 56
University of Rochester 7, 15, 71, 78, 141
upright napping procedure 42–3, 281

V
Valli, Katja 293–4
ventriloquist effect 43
verbal fluency 46–7, 111

Index

vestibular stimulation 216, 287
vibration, engineering dreams and 214–15, 217, 222, 223
virtual reality (VR) 210, 220–21, 285, 293, 303–4
virtual reality flying task 70–71, 285
visitation dreams 296–9, 304
visual artists, lucid dreaming and 277
visualisation 7, 9, 72, 158, 159, 177, 180, 186, 190, 192, 196, 203, 247, 252, 275

W

waking lives
  awakening from dreams *see* awakening
  bad dreams and *see* bad dreams
  concerns and surroundings, dreams reflecting 6, 7, 13, 25, 26, 35–6, 50, 60, 68, 259
  daydreams *see* daydreams
  dream skills and *see* dream skills
  feelings and *see* feelings
  lucid dreams and 176–8, 180–86, 188, 191, 192, 196, 226
  memory and *see* memory
  nightmares and 1–2, 3, 4–5, 103, 104, 106, 107, 108, 111, 112, 115, 117, 118, 119, 121, 122, 124, 125, 127, 130, 134, 138–40, 148–53, 158–60, 164, 166–9, 173, 174, 176–8, 180–83, 192, 193, 196, 197, 203, 208, 209, 210, 251, 252, 254
perception (dream) vs. imagination 39–41
physical health and *see* physical health
skill rehearsal in dreams 20, 21–4, 26, 50, 70, 72, 101, 119–20, 133, 270, 272, 284, 292–6, 300–302, 304, 305
wake-back-to-bed method 186–7, 190, 197, 199, 227–9
wake-induced lucid dreams 202–3
waking cognition 93–4
Wamsley, Erin 69
weak glimpse 48
wearables 204, 206, 214, 221–4, 280, 281
white dreams 48–9
Wii Fit balance task 70–71
witch dream 30–31

Z

Zadra, Antonio 242, 243
ZMax 222